17

Natural Products: The Secondary Metabolites

JAMES R. HANSON

University of Sussex

RS•C

ROYAL SOCIETY OF CHEMISTRY

Cover images © Murray Robertson/visual elements 1998–99, taken from the 109 Visual Elements Periodic Table, available at www.chemsoc.org/viselements

ISBN 0-85404-490-6

A catalogue record for this book is available from the British Library

Publlished by The Royal Society of Chemistry, Thomas Graham House, Science Park, Milton Road, Cambridge CB4 0WF, UK
Registered Charity No. 207890
For further information see our web site at www.rsc.org

Typeset in Great Britain by Alden Bookset, Northampton
Printed and bound in Italy by Rotolito Lombarda

Preface

Natural products are compounds produced by living organisms. They have played a significant role in the development of organic chemistry. The major chemical and physical methods of structure elucidation have been developed during the study of natural products. Many reactions of mechanistic importance have their origins in this area of chemistry. Natural products have provided challenging synthetic targets, and their biological activity has given leads for the development of valuable medicines.

The general metabolism that is common to all cells involves natural products that are known as the primary metabolites. On the other hand, there are natural products that are found in a limited range of species and indeed can give different species their distinctive characteristics. These are the secondary metabolites. The study of their structure and biosynthesis forms the subject of this book. There is a chemical logic in the determination of the structure of a natural product. It is the aim of this book to illustrate this chemical reasoning as applied to the different strategies that have been used for structure determination.

This book is aimed at the second-year undergraduate who has completed courses on functional group chemistry, stereochemistry and elementary spectroscopy. The elucidation of the structures of natural products brings many aspects of these together. The final chapter on biosynthesis is intended to act as a prelude to more advanced third-year courses on bio-organic chemistry. Since there is also a book on synthesis in this series, this aspect of natural product chemistry is not covered here. I have also tried to show how modern spectroscopic methods can provide information that in the past was obtained by chemical degradation.

I am indebted to Martyn Berry, Professor Alwyn Davies FRS, Professor Sir John Cornforth AC FRS, Professor John Mann and Dr Colin Drayton for their substantial help in the preparation of the manuscript.

James R. Hanson
University of Sussex

TUTORIAL CHEMISTRY TEXTS

EDITOR-IN-CHIEF

Professor E W Abel

EXECUTIVE EDITORS

Professor A G Davies
Professor D Phillips
Professor J D Woollins

EDUCATIONAL CONSULTANT

Mr M Berry

This series of books consists of short, single-topic or modular texts, concentrating on the fundamental areas of chemistry taught in undergraduate science courses. Each book provides a concise account of the basic principles underlying a given subject, embodying an independent-learning philosophy and including worked examples. The one topic, one book approach ensures that the series is adaptable to chemistry courses across a variety of institutions.

TITLES IN THE SERIES

Stereochemistry *D G Morris*
Reactions and Characterization of Solids
  S E Dann
Main Group Chemistry *W Henderson*
d- and f-Block Chemistry *C J Jones*
Structure and Bonding *J Barrett*
Functional Group Chemistry *J R Hanson*
Organotransition Metal Chemistry *A F Hill*
Heterocyclic Chemistry *M Sainsbury*
Atomic Structure and Periodicity *J Barrett*
Thermodynamics and Statistical Mechanics
  J M Seddon and J D Gale
Basic Atomic and Molecular Spectroscopy
  J M Hollas
Organic Synthetic Methods *J R Hanson*
Aromatic Chemistry *J D Hepworth,*
  D R Waring and M J Waring
Quantum Mechanics for Chemists
  D O Hayward
Peptides and Proteins *S Doonan*
Reaction Kinetics *M Robson Wright*
Natural Products: The Secondary
  Metabolites *J R Hanson*

FORTHCOMING TITLES

Mechanisms in Organic Reactions
Molecular Interactions
Lanthanide and Actinde Elements
Maths for Chemists
Bioinorganic Chemistry
Chemistry of Solid Surfaces
Biology for Chemists
Multi-element NMR
EPR Spectroscopy
Biophysical Chemistry

Further information about this series is available at www.rsc.org/tct

Order and enquiries should be sent to:

Sales and Customer Care, Royal Society of Chemistry, Thomas Graham House, Science Park, Milton Road, Cambridge CB4 0WF, UK

Tel: +44 1223 432360; Fax: +44 1223 426017; Email: sales@rsc.org

Contents

**1   The Classes of Natural Product
    and Their Isolation**   **1**

1.1   Introduction   1
1.2   The Classes of Secondary Metabolites   2
1.3   Polyketides and Fatty Acids   3
1.4   Terpenes and Steroids   6
1.5   Phenylpropanoids   15
1.6   Alkaloids   18
1.7   Antibiotics Derived from Amino Acids   22
1.8   Vitamins   22
1.9   Chemical Ecology   24
1.10 The Isolation of a Natural Product   28
1.11 The Stages in Structure Elucidation   30

**2   The Characterization and Determination of
    the Carbon Skeleton of a Natural Product**   **35**

2.1   The Characterization of a Natural Product   35
2.2   Spectroscopic Characterization   37
2.3   Simple Chemical Derivatives   45
2.4   The Determination of the Carbon Skeleton   46
2.5   Spectroscopic Methods in the Determination
      of the Carbon Skeleton   57

3   The Location of the Functional Groups
       and the Molecular Stereochemistry   63

3.1   Introduction   63
3.2   Spectroscopic Interrelationships   64
3.3   Chemical Methods   69
3.4   The Determination of Absolute Stereochemistry 75

4   Some Examples of Structure Elucidation   85

4.1   Santonin   85
4.2   Griseofulvin   89
4.3   Penicillin and Clavulanic Acid   93
4.4   Prostaglandins   96
4.5   Vitamin C   98

5   The Biosynthesis of Secondary Metabolites   105

5.1   Introduction   105
5.2   Biosynthetic Methodology   105
5.3   The Pathway of Carbon into Biosynthesis   107
5.4   The Biosynthesis of Polyketides   108
5.5   Terpenoid Biosynthesis   112
5.6   The Biosynthesis of Phenylpropanoids   119
5.7   Alkaloid Biosynthesis   121
5.8   Other Natural Products Derived from Amino Acids 124

Further Reading   131
Answers to Problems   135
Subject Index   145

1

The Classes of Natural Product and Their Isolation

Aims

The aims of this chapter are to introduce the main classes of natural product and to show how they may be isolated. By the end of this chapter you should understand:

- The motivation for examining the structures of natural products
- The distinction between primary and secondary metabolites
- The structural characteristics of the major classes of natural product and recognize their biosynthetic building blocks
- The chemistry underlying the methods of isolating natural products
- The stages in the elucidation of the structure of a natural product

1.1 Introduction

Natural products are organic compounds that are formed by living systems. The elucidation of their structures and their chemistry, synthesis and biosynthesis are major areas of organic chemistry. Naturally occurring compounds may be divided into three broad categories. Firstly, there are those compounds which occur in all cells and play a central role in the metabolism and reproduction of those cells. These compounds include the nucleic acids and the common amino acids and sugars. They are known as **primary metabolites**. Secondly, there are the high-molecular-weight polymeric materials such as cellulose, the lignins and the proteins which form the cellular structures. Finally, there are those compounds that are characteristic of a limited range of species. These are the **secondary metabolites**. Most primary metabolites exert their biological effect within the cell or organism that is responsible for their production. Secondary metabolites, on the other hand, have often attracted interest because of their biological effect on other organisms.

The biologically active constituents of medicinal, commercial and poisonous plants have been studied throughout the development of organic chemistry. Many of these compounds are secondary metabolites. It has been estimated that over 40% of medicines have their origins in these natural products. A number of screening programmes for bioactive compounds exist and have led to new drugs, for example taxol, which is used for the treatment of various cancers. Natural products often have an ecological role in regulating the interactions between plants, micro-organisms, insects and animals. They can be defensive substances, anti-feedants, attractants and pheromones. Chemotaxonomy provides another reason for examining the constituents of plants. Phytochemical surveys can reveal natural products that are "markers" for botanical and evolutionary relationships.

> A pheromone is a substance released by one individual of a species which causes a change in behaviour by another member of the same species.

> Chemotaxonomy involves the use of natural products in the classification of species.

> A phytochemical is a natural product produced by a plant.

1.2   The Classes of Secondary Metabolites

At first sight, the structures of secondary metabolites may seem to be bewilderingly diverse. However, the majority of these compounds belong to one of a number of families, each of which have particular structural characteristics arising from the way in which they are built up in nature, *i.e.* from their biosynthesis. The classes of secondary metabolites are:

- Polyketides and fatty acids
- Terpenoids and steroids
- Phenylpropanoids
- Alkaloids
- Specialized amino acids and peptides
- Specialized carbohydrates

> The polyketide chain is assembled from acetate units:

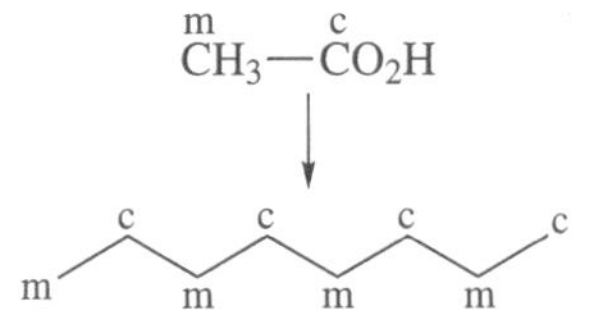

$$\overset{m}{C}H_3 \!-\! \overset{c}{C}O_2H$$

> Isoprene is 3-methylbuta-1,3-diene. Some simple terpenes give isoprene on pyrolysis.

> Phenylpropanoids have the carbon skeleton:

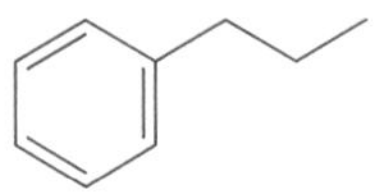

Polyketides are formed by the linear combination of **acetate** (ethanoate) units derived from the "building block" acetyl co-enzyme A. **Terpenoids** and **steroids** are assembled in nature from **isoprenoid** C_5 units derived from isopentenyl (3-methylbut-3-en-1-yl) pyrophosphate. These C_5 units are linked together in a head-to-tail manner. They have a characteristic branched chain structure. A further group of natural products are those containing a **phenylpropanoid** (C_6–C_3) unit.

The **amino acids** are the building blocks for peptides and proteins. Although the amino acids are normally considered as primary metabolites, there are some unusual amino acids that are of restricted occurrence. Some antibiotics such as the penicillins are formed from small peptides. The **alkaloids** are a structurally diverse group of natural products containing nitrogen. The nitrogenous portions of the alkaloids are derived from amino acids such as ornithine, lysine, tyrosine or tryptophan.

Although **sugars (carbohydrates)** such as glucose are typical primary metabolites, there are other sugars that are of a much more limited occurrence. Some of these less common sugars are attached to natural products as part of a **glycoside**. The non-sugar portion is known as the **aglycone**, and may be a terpenoid, alkaloid or polyketide.

1.3   Polyketides and Fatty Acids

Polyketides are natural products that are formed by the stepwise condensation of acetate (ethanoate) units. In the resultant carbon chain, alternate carbon atoms come from the methyl and carboxyl groups of the acetate building block. The acetate origin of these compounds leads to a preponderance of even-numbered carbon chains. Their biosynthesis is discussed in Chapter 5. Many plant oils and animal fats contain long-chain monocarboxylic acids known as fatty acids. In the fatty acids, the carbonyl group of the acetate units is reduced during the course of the chain assembly process. Dehydrogenation and oxidative processes may subsequently give the unsaturated fatty acids. If some of the carbonyl groups are not removed during the biosynthesis, various internal aldol-type condensations may occur, leading after dehydration and enolization to aromatic compounds. Other cyclizations may give oxygen heterocycles. A characteristic feature of these compounds is oxygenation on alternate carbon atoms.

Fatty acids are found in fats both as the free acids and, more often, combined as esters with alcohols such as glycerol and cholesterol. The common fatty acids have an even number of carbon atoms, typically C_{12}–C_{20}, linked together in a straight chain with up to four double bonds. The double bonds usually possess the *cis* geometry and are not conjugated. In plants the fatty acids and the corresponding alcohols are found in leaf waxes and seed coatings. Myristic acid (C_{14}) is found in nutmeg seeds and palmitic acid (C_{16}) is found in almost all plant oils. Stearic acid (C_{18}) occurs in large amounts in animal fat. These exemplify the common occurrence of even-numbered carbon chains in fatty acids. The sodium salts of the fatty acids are used as soaps.

Unsaturated fatty acids are important to us in food. Oleic acid (*cis*-octadec-9-enoic acid, **1.1**) is the most widely distributed, and is a major constituent of olive oil. Linoleic (**1.2**) and linolenic (**1.3**) are more highly unsaturated and are found in linseed oil. Linolenic acid is easily oxidized by air, and is one of the "drying oils" used in paints and varnishes. The methylene which lies between the two double bonds is very sensitive to radical reactions. Linolenic acid is oxidized by plants to jasmonic acid

The aldol condensation between a carbonyl-activated carbanion and another carbonyl group gives a β-hydroxy ketone. The condensation followed by enolization of polycarbonyl compounds creates a phenol:

(**1.4**), which is a signalling substance that stimulates plant defence mechanisms. The unsaturated C_{20} arachidonic acid (**1.5**) is a precursor of the **prostaglandin** hormones (see Chapter 5).

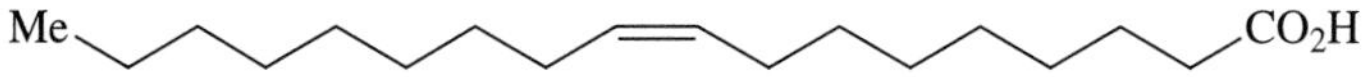
Me
CO_2H

1.1

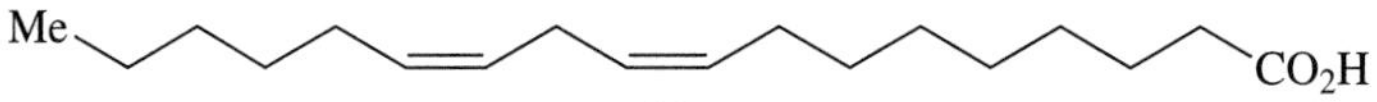
Me
CO_2H

1.2

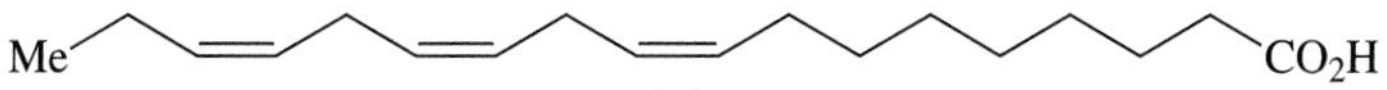
Me
CO_2H

1.3

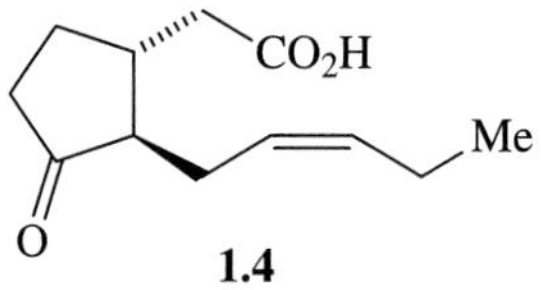
CO_2H
Me
O

1.4

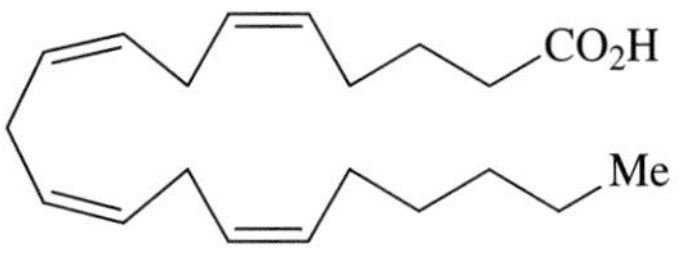
CO_2H
Me

1.5

The *Basidiomycetes* are fungi with large fruiting bodies. They include the mushrooms, toadstools and bracket fungi found growing on wood.

Some highly unsaturated **polyacetylenes** derived from polyketides are found in plants of the Dahlia family and in Basidiomycete fungi. The antibiotic mycomycin (**1.6**) is a polyacetylene which also contains an unusual allene unit. The antifungal agent wyerone (**1.7**) is produced by the broad bean (*Vicia fabae*) when it is attacked by micro-organisms.

$$HC{\equiv}C{-}C{\equiv}C{-}CH{=}C{=}CH{-}\overset{cis}{CH{=}CH}{-}\overset{trans}{CH{=}CH}{-}CH_2CO_2H$$

1.6

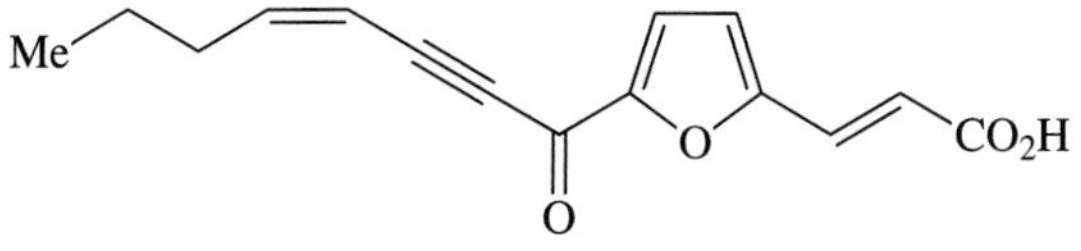
Me
O
O
CO_2H

1.7

A lactone is a cyclic ester formed from an alcohol and an acid group in the same molecule.

Large ring lactones, known as **macrolides**, are formed by some micro-organisms. The macrolides include some important antibiotics such as erythromycin (**1.8**), which is produced by a Streptomycete. Other macrolides, such as the avermectins and milbemycins, are produced commercially for the treatment of parasites in animals and as insecticides.

Many aromatic natural products with oxygen atoms attached to alternate carbon atoms are polyketides. These compounds are grouped by the number of their constituent acetate units into tri-, tetra-, penta-, hexa- and heptaketides. Some typical examples from micro-organisms are the tetraketide 6-methylsalicyclic acid (**1.9**) from *Penicillium patulum*, the pentaketide mellein (**1.10**) from *Aspergillus melleus*, the heptaketide antibiotic griseofulvin (**1.11**) from *P. griseofulvum* and mycophenolic acid (**1.12**) from *P. brevicompactum*. Griseofulvin (see Chapter 4) has been used to treat fungal infections of the skin and mycophenolic acid is an immunosuppressive agent.

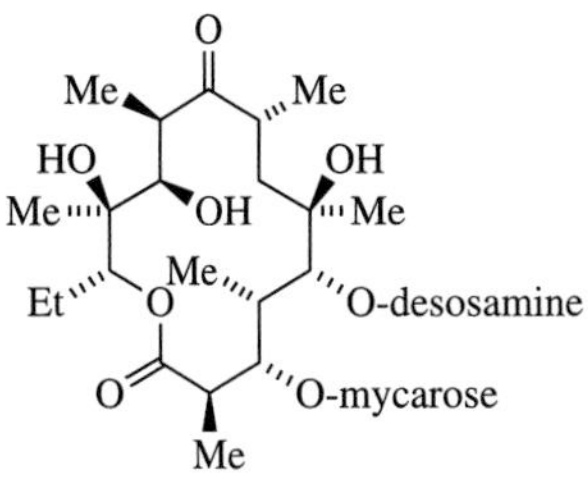

1.8

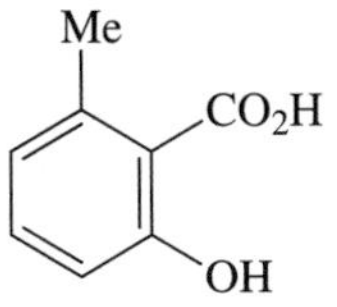

1.9 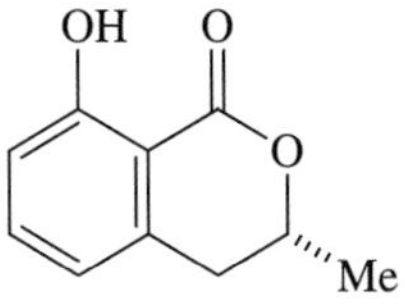   **1.10**

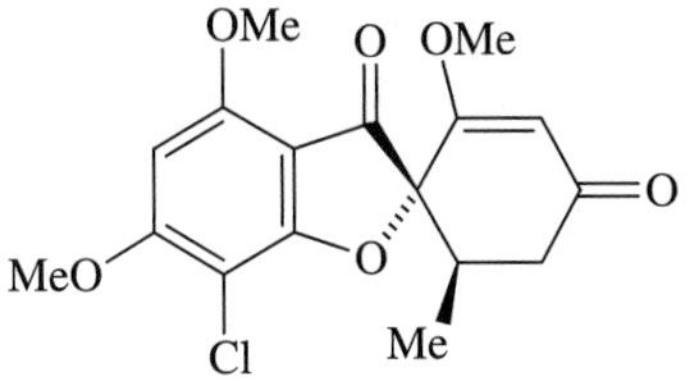

1.11 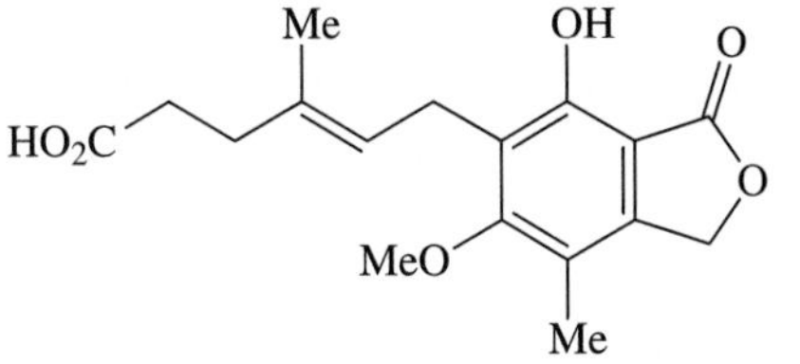   **1.12**

The aromatic ring may be cleaved to form compounds such as patulin (**1.13**) from *P. patulum*. The aflatoxins (*e.g.* **1.14**) from *A. flavus* are among the most dangerous fungal toxins and cause liver cancer. Their identification in 1962, as the causative agents of liver cancer in Christmas turkeys being fattened on groundnuts which were infected with *A. flavus*, was a tale of chemical detective work and careful structure elucidation. The tetracycline antibiotics (*e.g.* tetracycline, **1.15**) are valuable antibiotics which are produced on a large scale from *Streptomyces aureofaciens*.

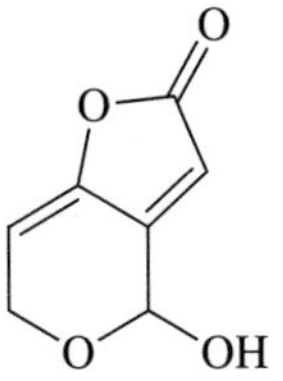

1.13 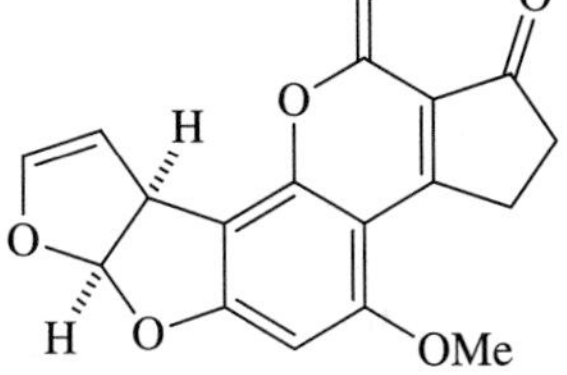   **1.14**

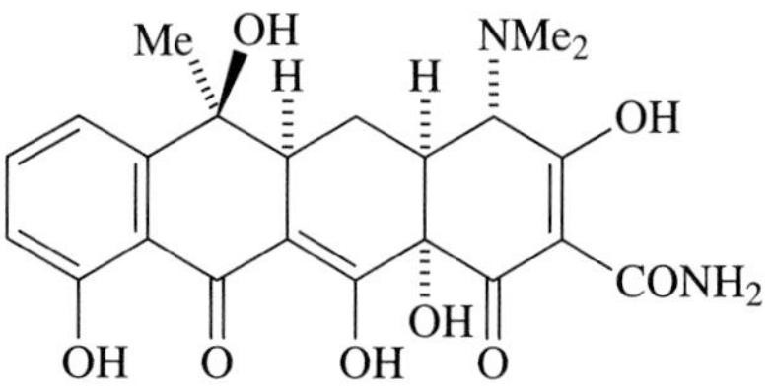

1.15

Worked Problem 1.1

Q   Identify the acetate units in the lichen product eugenitin (**1**).

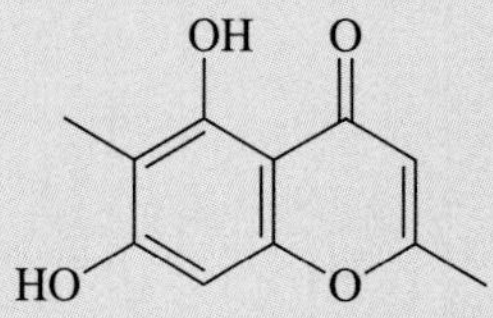

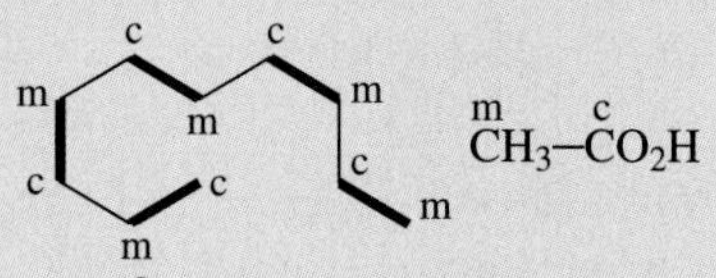

A   This natural product is a pentaketide possessing oxygen functions on alternate carbon atoms (see **2**). These act as markers for the carbonyl group of the acetate unit. Note that there is an extra carbon atom which has probably come from methionine.

1.4   Terpenes and Steroids

The terpenes are compounds that are built up from **isoprene** units. Their structures are divisible into the C_5 isoprene units (C—C—C—C) linked

$$\underset{\qquad\;\;\;|}{\qquad\;\;\;C}$$

in a head-to-tail manner. This isoprene rule, developed by Ruzicka in 1921, provided a useful guide in structure determination. The origin of the C_5 unit (isopentenyl pyrophosphate, **1.16**) is discussed in Chapter 5.

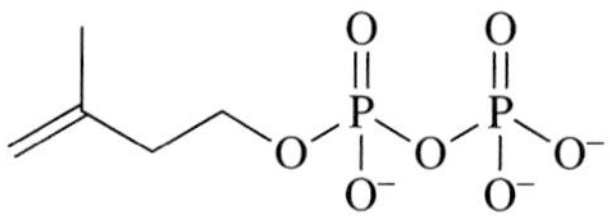

1.16

The terpenes are classified by the number of these C_5 isoprene units that they contain. The classes are:

- Monoterpenoids, C_{10}
- Sesquiterpenoids, C_{15}
- Diterpenoids, C_{20}

- Sesterterpenoids, C_{25}
- Triterpenoids, C_{30}
- Carotenoids, C_{40}

Natural rubber is a polyisoprenoid substance. The steroids are derived from the tetracyclic triterpenoids. Isoprene units are sometimes found as components of other natural products.

Most of the terpenes have cyclic structures. The majority of the terpenoid cyclizations which take place in living systems are of an acid-catalysed type. The branched chain nature of the isoprenoid backbone coupled with readily protonated functional groups (*e.g.* alkenes) lend themselves to acid-catalysed rearrangements during the course of these biosynthetic reactions. Consequently, there are some terpenoid skeletons that do not appear, on first inspection, to be derived from the regular combination of isoprenoid units.

Monoterpenoids are major components of the aromas of plants. These volatile natural products, known as **essential oils**, form the basis of the perfumery and flavouring industries. Distillation of these oils meant that it was possible to obtain quantities of these terpenes for structural studies. The structures of many of the simple monoterpenes were established between 1890 and 1920 through the work of Wallach, Wagner, Tiemann, Semmler and Perkin. Although modern gas chromatographic analyses of these oils often shows them to be highly complex mixtures of terpenoid and non-terpenoid natural products, one or two monoterpenes usually predominate. Thus geraniol (**1.17**) is a major component of geranium oil (*Pelargonium graveolens*) and its isomer, linalool (**1.18**), is found in the oil of a garden herb, clary sage. Both enantiomers of linalool occur naturally. Citral (**1.19**), a constituent of lemon oil, is obtained commercially from lemon grass oil (*Cymbopogon flexuosus*). Herbs, including the mints, the sages and rosemary, are the source of many terpenes. Menthol (**1.20**) is found in the essential oil of the field mint, *Mentha arvensis*, and possesses useful physiological properties including local anaesthetic and refreshing effects. It is used to flavour sweets, tobacco and toothpaste. Terpineol (**1.21**) and α-pinene (**1.22**) are found in pine oil (turpentine). Camphor (**1.23**), which was isolated from the camphor tree, *Cinnamomum camphora*, but is now made commercially from α-pinene, is used to protect clothes from moths.

1.17   1.18   1.19

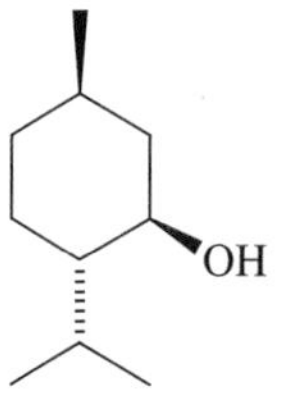

1.20

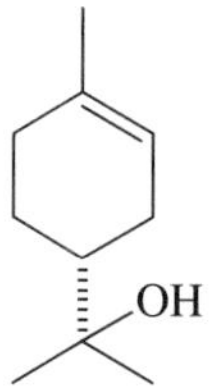

1.21

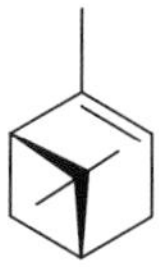

1.22

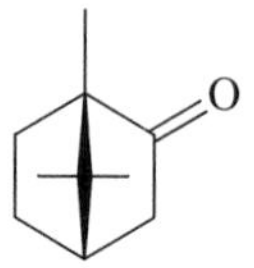

1.23

Amongst the more highly oxygenated monoterpenoids are a family of cyclopentanes known as the **iridoids**. Some compounds of this series such as iridodial (**1.24**) are found in ants. Iridodial is in equilibrium with its hemiacetal. The corresponding lactone, nepetalactone, obtained from catmint (*Nepeta cataria*), is an attractant for cats. The seco-iridoids such as seco-loganin (**1.25**) are precursors of the terpenoid portion of the indole alkaloids (see Chapter 5).

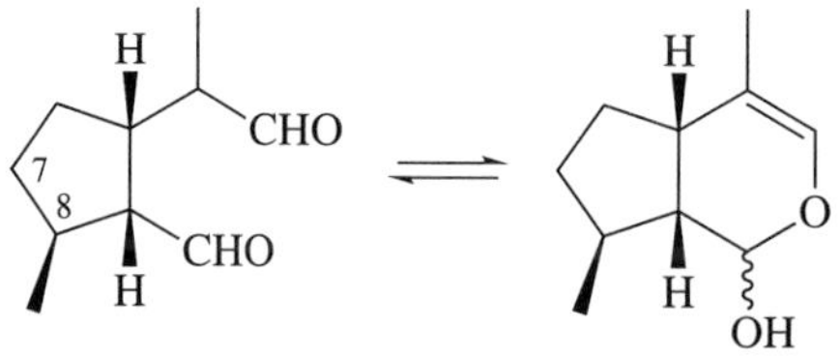

1.24

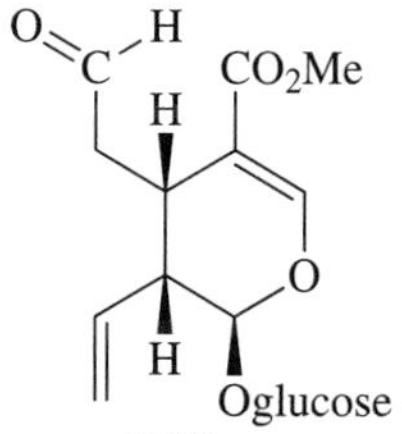

1.25

Some **sesquiterpenes** are found in the higher boiling portions of essential oils. These include caryophyllene (**1.26**) from oil of cloves, humulene (**1.27**) from oil of hops, cedrene (**1.28**) from cedar wood oil and longifolene (**1.29**) from Indian turpentine oil (*Pinus ponderosa*). These structures illustrate the diversity of the carbon skeletons found among the sesquiterpenes. Sesquiterpenoid lactones are common biologically active constituents of plants of the Compositae family. Santonin (**1.30**) from *Artemisia maritima* (wormwood) was used in medicine for the elimination of intestinal worms. The evidence for the structure of santonin is described in Chapter 4. Derivatives of the Chinese drug qinghaosu, artemisinin (**1.31**), obtained from *Artemisia annua*, have recently been recommended by the World Health Organization for the treatment of resistant strains of malaria. This compound contains an unusual peroxide which is associated with its biological activity. The plant hormone abscisic acid (**1.32**), which stimulates leaf fall and dormancy in plants, was identified in 1965 and shown to be a sesquiterpenoid.

The *Compositae* are the largest family of flowering plants. Their flowers have characteristically a large number of petals. Typical examples are the daisy and the sunflower.

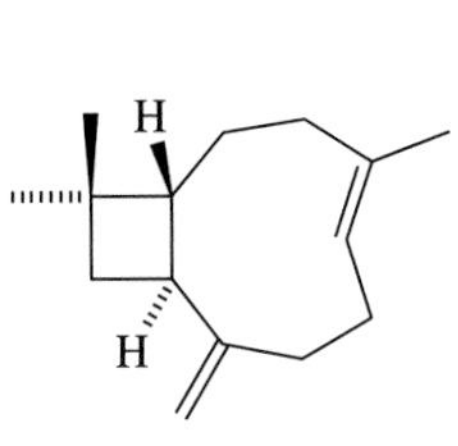

1.26 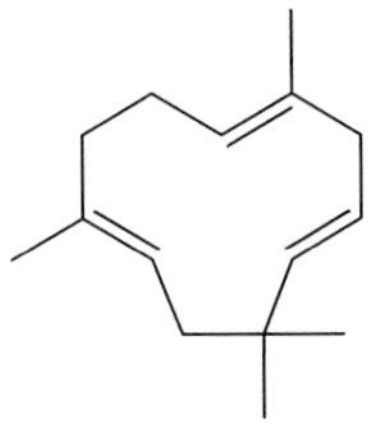   1.27 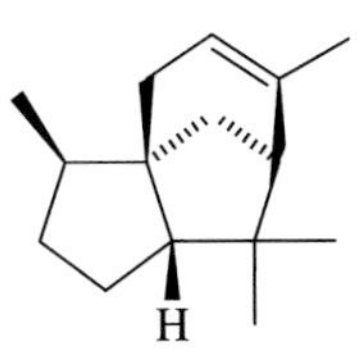   1.28

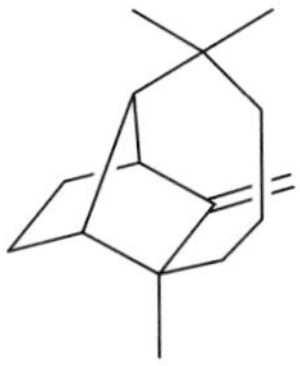

1.29 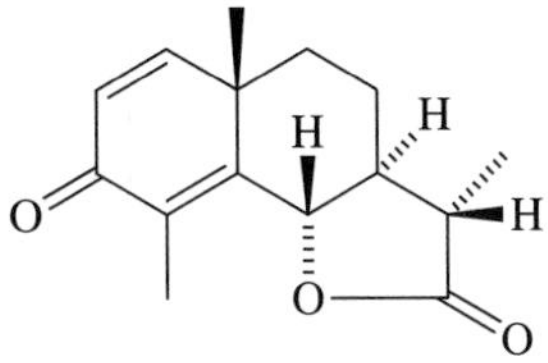   1.30 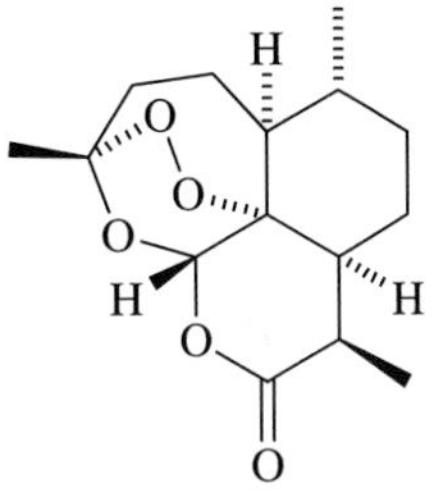   1.31 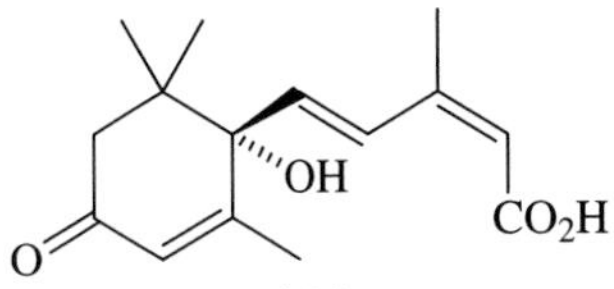   1.32

A number of fungal metabolites are sesquiterpenoids. Some like the botrylanes (*e.g.* botrydial, **1.33**), which are produced by the phytopathogenic fungus *Botrydis cinerea*, are phytotoxins and others like the trichothecenes (*e.g.* **1.34**), which are characteristic metabolites of *Fusarium* species, are toxic to mammals.

A phytotoxin is a substance that is toxic to plants.

Botrytis cinerea is a grey powdery mould that is sometimes found on lettuces and tomatoes. *Fusarium* species are widespread fungi, many of which are pathogenic to plants. A few also affect insects and man.

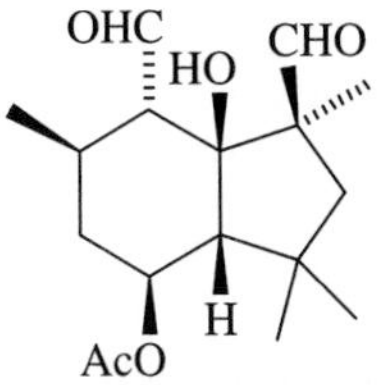

1.33 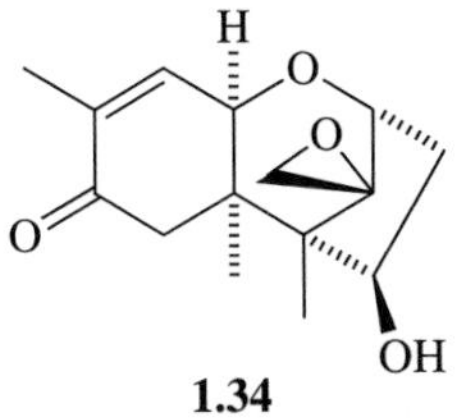   1.34

Many of the **diterpenoids** are wood resin products. Abietic acid (**1.35**) is the major component of colophony, the wood rosin obtained from *Pinus* and *Abies* species. It is used in varnishes and resin soaps. Although crude samples of the resin acids were isolated in the 19th century, there was considerable confusion concerning their identity. The carbon skeleton of abietic acid was not fully established until the work of Ruzicka in the 1930s, and the stereochemistry was not clarified until 1948 by Barton. Other resin acids include pimaric acid (from *Pinus* species) and podocarpic acid (**1.36**) from *Podocarpus cupressinum*. Neutral constituents of wood resins include manoyl oxide (**1.37**). The acids and phenols provide protection for the wood against fungal and insect attack. More highly oxygenated diterpenoids possessing the clerodane skeleton are insect anti-feedants, and are found in the leaves of, for example, *Ajuga*,

These plants belong to the *Lamiaceae* (also known as the *Labiatae*). This family of plants include many common garden herbs such as mint, rosemary and sage. Bugleweed is an *Ajuga* species and the *Salvias* are common decorative garden flowers.

Salvia, *Scutellaria* and *Teucrium* species. The orange and red pigments of the leaves of the decorative *Coleus* species (*e.g.* coleon B) and the active principles of a Chinese drug, Tan-shen (*Salvia miltiorrhiza*), are diterpenoid quinones.

1.35   **1.36**   **1.37**

The prefix "*ent*" is used to refer to "enantiomer of", in this case kaurene.

The fungus *Gibberella fujikuroi* is a serious pathogen on rice, causing the plants to grow rapidly and then die.

Stevia rebaudiana is a shrub which belongs to the *Compositae* family. It is grown commercially in South America, Korea and Japan.

Plants of the spurge family belong to the *Euphorbiaceae*. A number produce an irritant latex. Poinsettias are Euphorbias.

Diterpenes occur in both enantiomeric series. Diterpenes with the *ent*-kaurene skeleton are widespread. *ent*-Kaurenoic acid (**1.38**, R = H) is a biosynthetic precursor of the diterpenoid **gibberellin plant growth hormones**. The best known of these plant hormones, gibberellic acid (**1.39**), is also produced in large amounts as a phytotoxin by the fungus *Gibberella fujikuroi*. It is produced commercially and used in the malting step in beer manufacture where it stimulates the production of the starch hydrolytic enzyme, α-amylase. It is also used in the production of seedless grapes. Steviol (**1.38**, R = OH) is the aglycone of the natural sweetener stevioside, which is obtained commercially from the plant *Stevia rebaudiana*. Stevioside is used in some countries as a non-nutritive sweetener in low-calorie drinks. A number of diterpenoids possess antitumour activity. One of these, taxol (or paclitaxel, **1.40**), was originally obtained from the bark of the Pacific yew, *Taxus brevifolia*, but it is now made semi-synthetically from more readily available taxanes. It possesses powerful activity against a number of tumours and it is used in the treatment of breast and ovarian cancer. On the other hand, some of the diterpenoid constituents of *Euphorbia* species have powerful skin irritant and co-carcinogenic properties.

1.38   **1.39**

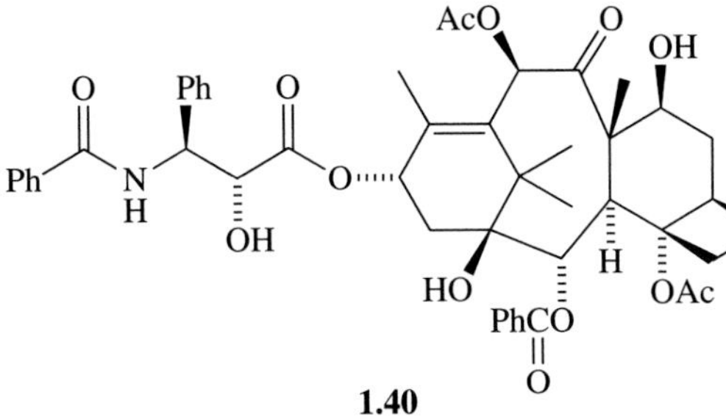

1.40

The simplest **triterpene**, squalene (**1.41**), was first isolated from fish liver oils. Subsequently, it has been found in plant oils and mammalian fats. The common tetracyclic triterpene lanosterol (**1.42**) is a major constituent of wool fat and its esters are found in lanolin cream. Other triterpenes, such as the α- and β-amyrins (**1.43**), are found in wood resin and the bark of many trees. A triterpene lactone, abietospiran, crystallizes on the surface of the bark of the silver fir, *Abies alba*, giving it a grey-white appearance. Glycyrrhetinic acid is a triterpene found in liquorice, and has healing properties in the treatment of peptic ulcers.

Liquorice is obtained from the roots of *Glycyrrhiza glabra*.

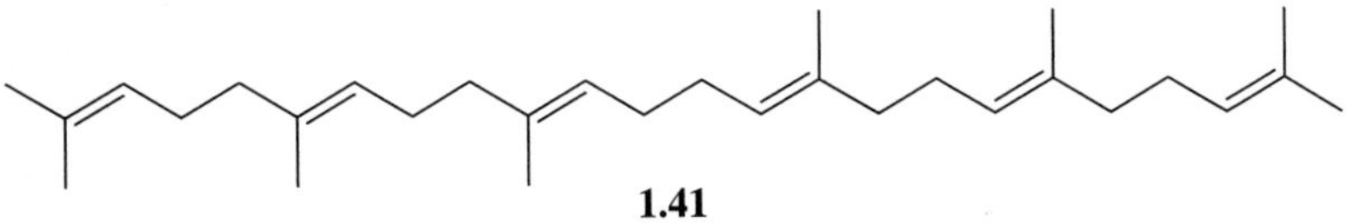

1.41

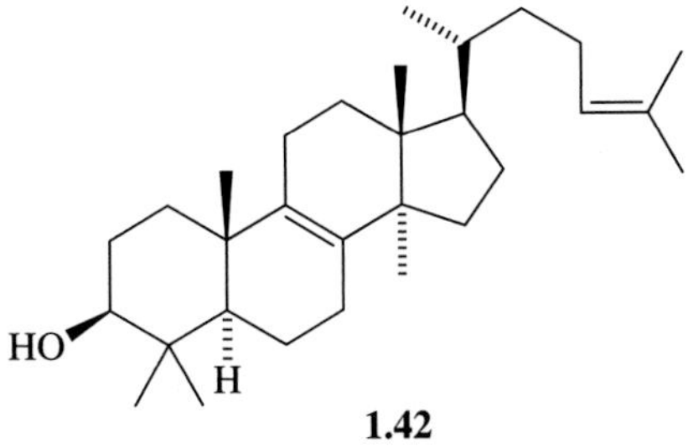

1.42

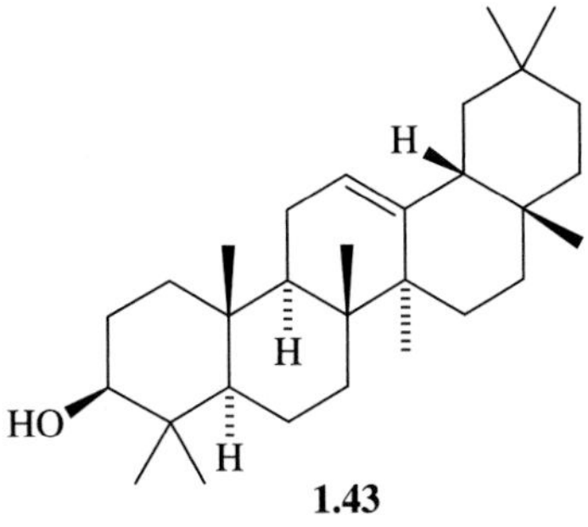

1.43

Triterpenes have been found in the fruiting bodies of a number of fungi, particularly from *Basidiomycetes* such as *Polyporus* (polyporenic acids) and *Ganoderma* (ganoderic acids) species. The traditional Chinese medicine ginseng (*Panax ginseng*) contains glycosides of triterpenes such as protopanaxadiol. Recently, derivatives of betulin (from the bark of the beech tree) have shown interesting activity against the human immuno-deficiency virus.

More highly oxidized and degraded triterpenes are exemplified by limonin (**1.44**), which is a bitter principle obtained from lemon and orange seeds. Although limonin was first isolated by Bernay in 1841, its structure was not finally established until 1960 as the result of a collaborative effort involving work by Barton in the UK, Jeger and Arigoni in Switzerland, and Corey in the USA. Other members of this group have been obtained from the heartwood of trees in the *Meliaceae* and *Rutaceae* families. One of these, azadirachtin, from the Neem tree, has strong insect anti-feedant activity against locusts. Its structure was finally established in 1987 by the work of three groups, those of Ley, Nakanishi and Kraus.

The *Meliacae* are a family of trees that provide a number of the tropical hardwoods which are used in furniture.

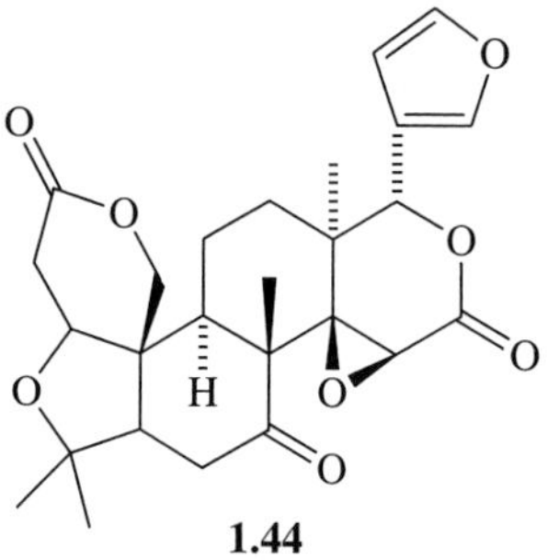

1.44

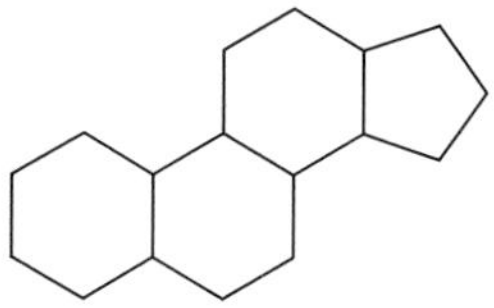

1.45

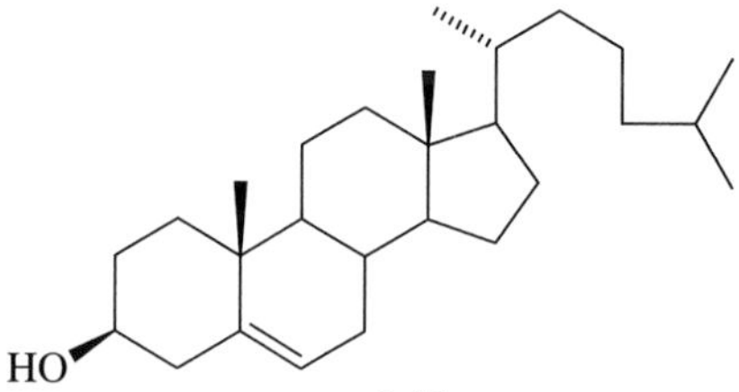

1.46

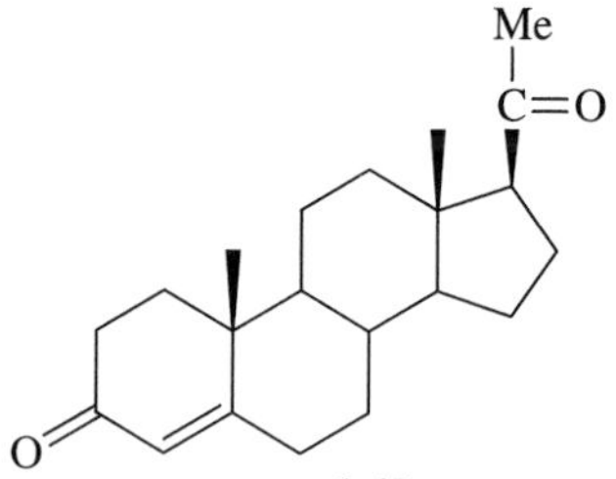

1.47

The **steroids** are derived from tetracyclic triterpenes and possess a cyclopentaperhydrophenanthrene backbone (**1.45**). The availability, crystallinity and well-defined conformation of the steroids have meant that they have become suitable substrates with which to investigate the influence of steric factors on reaction rates and mechanisms. Cholesterol (**1.46**) forms an important constituent of lipid membranes, and the more highly degraded steroids include the **steroid hormones**. Cholesterol is a typical mammalian sterol, whereas ergosterol is found in fungi and stigmasterol in plant oils. Among the steroid hormones, the progestogens (*e.g.* progesterone, **1.47**) and the estrogens (*e.g.* estradiol, **1.48**) are female hormones responsible for female sexual characteristics, for the maintenance of pregnancy and for the control of the menstrual cycle. Modified estrogens and progestogens are used in the oral contraceptive. Some breast cancers are estrogen dependent. The enzyme system aromatase, which catalyses the conversion of testosterone (**1.49**) to estradiol, is a target for cancer chemotherapy. Androgens such as testosterone are male sexual hormones. Among other properties, they have a stimulating effect on the development of muscle (anabolic effect). Anabolic steroids have been used in stimulating the growth of beef cattle. The **cortical steroids** (*e.g.* cortisone, **1.50**) are produced by the adrenal cortex. They have two main functions. The mineralcorticoids (*e.g.* aldosterone) regulate mineral balance, and the glucocorticoids promote the conversion of protein to carbohydrate (gluconeogenesis) and its storage as glycogen. Cortical steroids have an immunosuppressive activity and reduce inflammation. They are used in the treatment of rheumatoid arthritis, asthma and in creams for reducing inflammation. The **vitamin D** metabolite 1α,25-dihydroxyvitamin D$_3$ (**1.51**) controls gastrointestinal calcium and phosphate absorption and promotes the mineralization of bones.

Nandrolone is an anabolic steroid that is abused by some athletes.

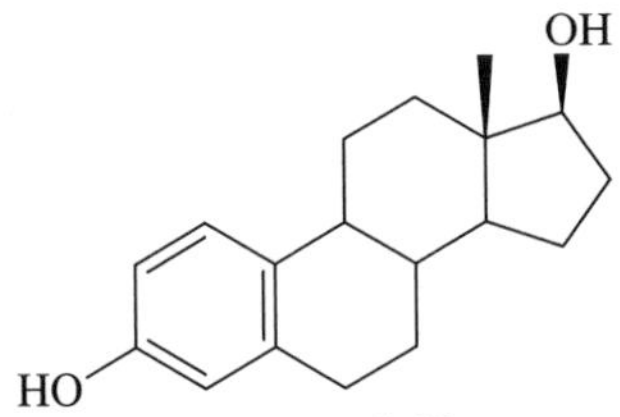

1.48

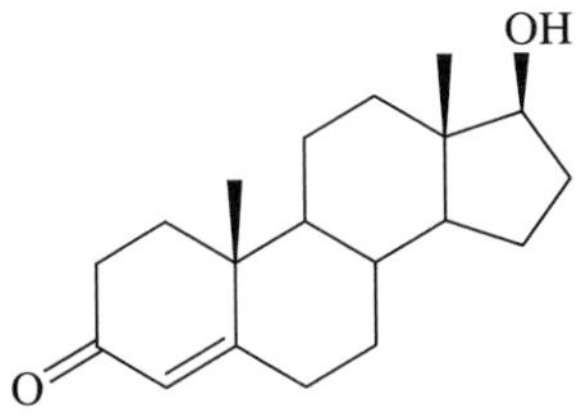

1.49

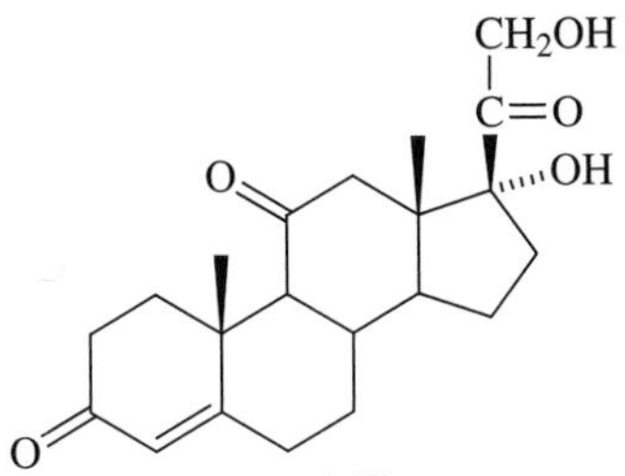

1.50

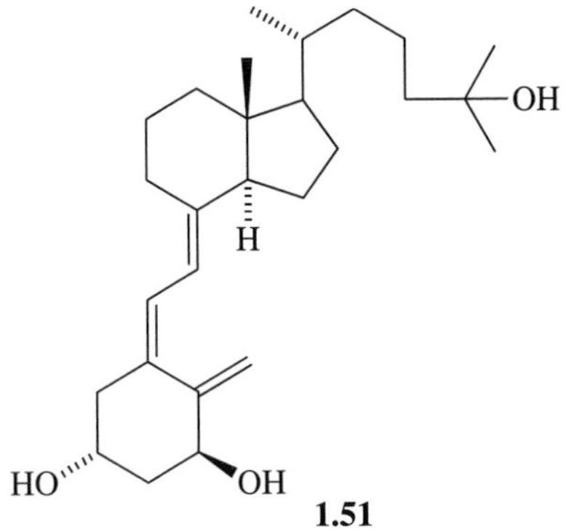

1.51

Many sterols occur as glycosides typified by the steroidal **saponins**. These are responsible for the foaming produced by many plants. Diosgenin (**1.52**), a steroidal sapogenin from the Mexican yam, and hecogenin, from sisal, are used as starting materials for the partial synthesis of the steroidal hormones. The steroidal alkaloids, such as solasodine, occur in plants of the *Solanaceae*, including the tomato and potato. A number of plant steroids possess a useful pharmacological activity. These include the digitalis glycosides (cardenolides) from the foxglove, *Digitalis lanata*. These are used in the treatment of heart failure.

Other steroids, such as the **ecdysteroids**, are insect hormones whilst the **brassinosteroids** are plant hormones.

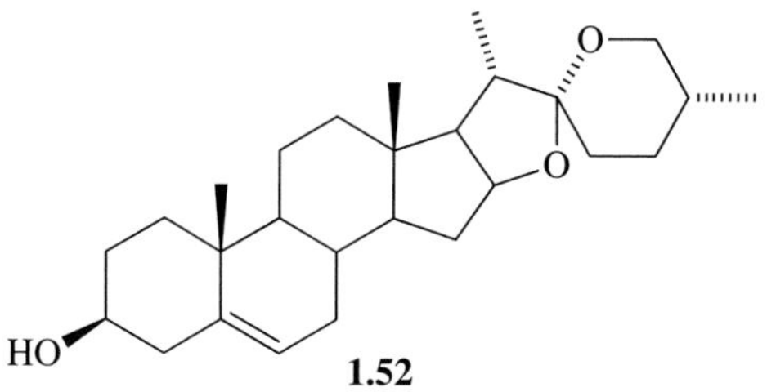

1.52

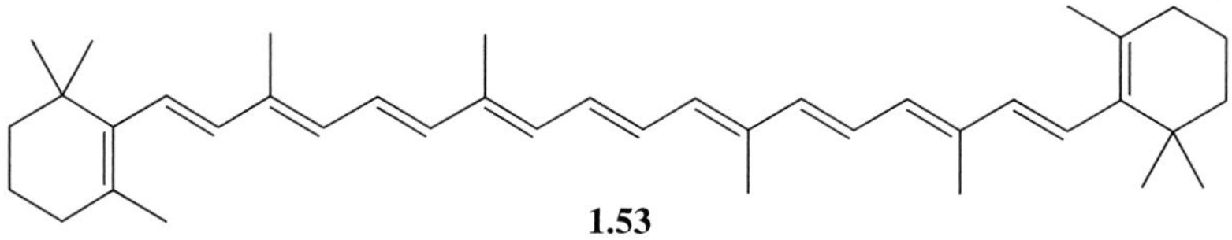

1.53

Some carotenoids are produced commercially as colouring matters for foods.

The **carotenoids** are red or yellow pigments that are found in many plants. Thus β-carotene (**1.53**) provides the red colouring matter of carrots and lycopene is the deep-red pigment of tomatoes. The carotenoids are important as precursors of **vitamin A**, which plays a central role in vision. The carotenoids are good anti-oxidants and contribute beneficial effects to many foods.

Worked Problem 1.2

Q Identify the isoprene units in paniculide B (**3**), a product of tissue culture from *Andrographis paniculata*.

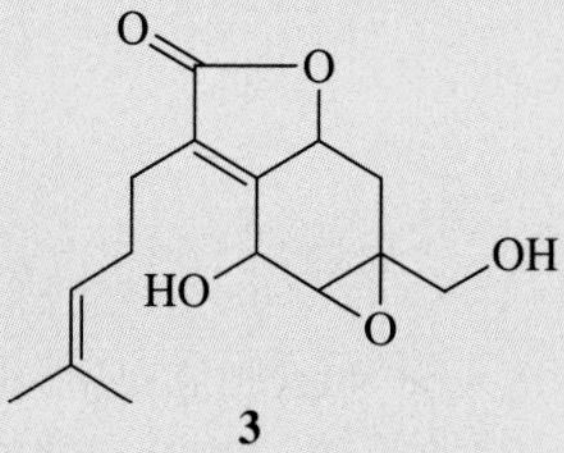

3 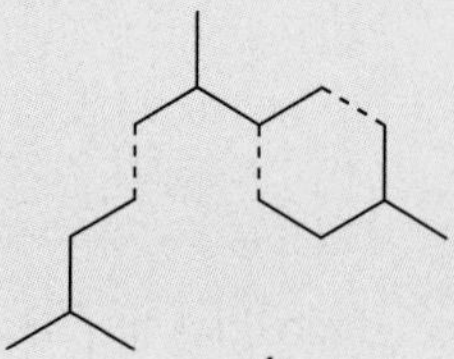   **4** 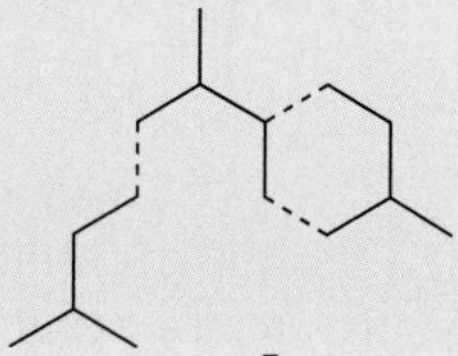   **5**

A The isoprene units are show in **4** and **5**. There are two ways in which the six-membered ring might be dissected into isoprene units. A biosynthetic experiment has been carried out using $[1,2\text{-}^{13}C_2]$acetate to show that folding as in **4** occurs. Return to this example when you have read Chapter 5 to see how the labelling pattern from $[1,2\text{-}^{13}C_2]$acetate would provide this distinction.

1.5   Phenylpropanoids

The recognition that many naturally occurring aromatic compounds possessed a three-carbon chain attached to the ring (see **1.54**) led to them being grouped together as the phenylpropanoids. Their biosynthetic origin via shikimic acid (**1.55**) is discussed in Chapter 5. Aromatic compounds that are biosynthesized by this route can be distinguished by their oxygenation pattern from those that are of polyketide origin. Hydroxylation occurs at the position *para* to the C_3 chain and then in the *meta* position, rather than on alternate carbon atoms (see, for example, **1.57**). The structures of a number of simple phenylpropanoids, which are widespread plant products, were established in the late 19th century. Cinnamic acid (**1.56**, R = H), 4-hydroxycinnamic acid (coumaric acid, **1.56**, R = OH), 3,4-dihydroxycinnamic acid (caffeic acid) and their methyl ethers (*e.g.* ferulic acid) are found in the free state and as their esters. Neutral compounds such as eugenol (**1.57**) (an aromatic constituent of clove oil) are components of essential oils. The important aromatic amino acids phenylalanine, tyrosine and tryptophan are formed by this route. The cyclization of the *Z*-isomers of the alkenes leads to **coumarins** such as umbelliferone (**1.58**). Some of these compounds are associated with the photosensitizing properties of plants of the *Umbelliferae*. Oxidative coupling of two phenylpropanoid units, such as coniferyl alcohol (**1.59**), leads to the **lignan** carbon skeleton, exemplified by pinoresinol (**1.60**). Podophyllotoxin (**1.61**), from the Hindu drug *Podophyllum emodi*, is an interesting cytotoxic agent.

Eugenol can be isolated from cloves by steam distillation.

The *Umbelliferae* are a large family in which the flowers are grouped in a characteristic "umbrella" shape. This family includes hogweed, carrots and parsnips

Cytotoxic = toxic to cells.

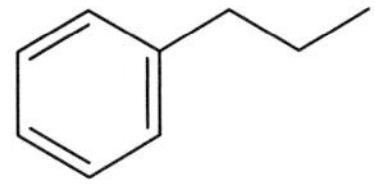

1.54 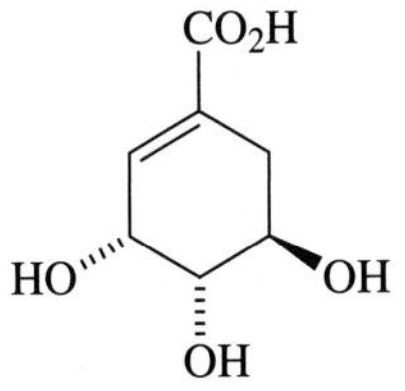   **1.55** 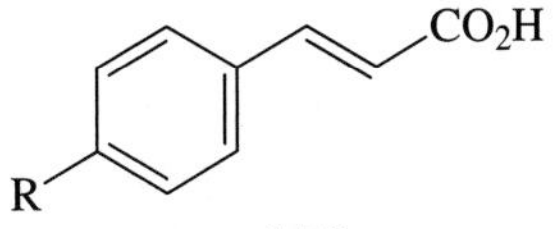   **1.56**

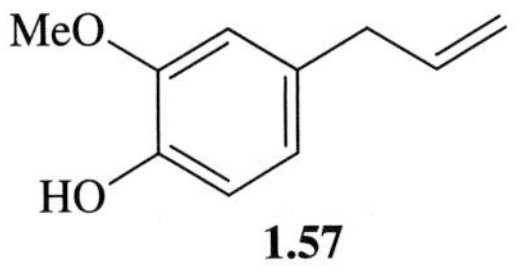

1.57 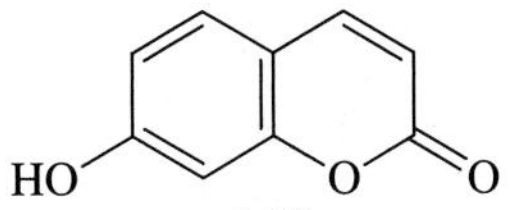   **1.58** 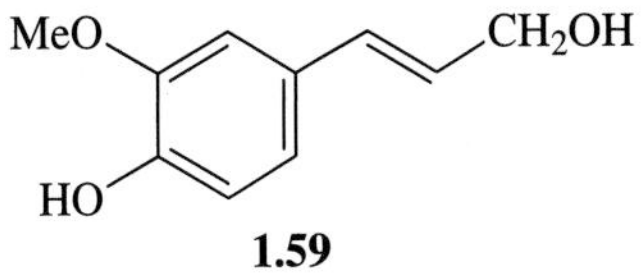   **1.59**

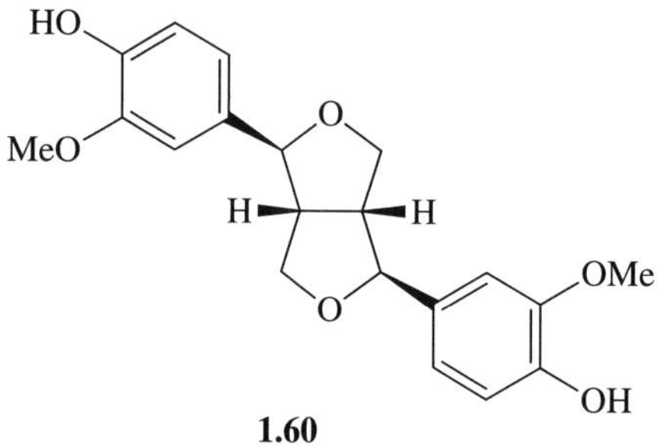

1.60 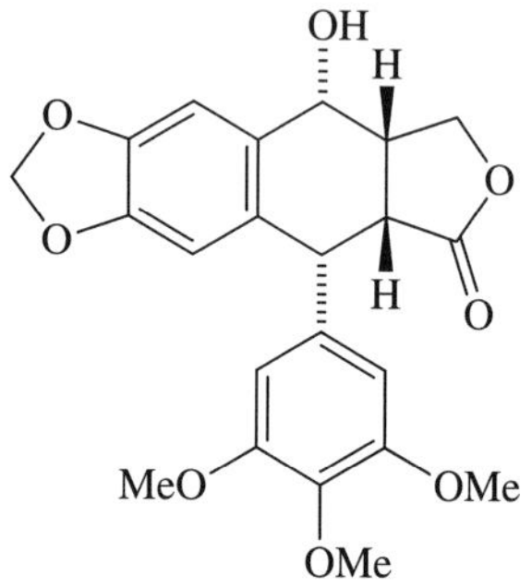   **1.61**

4-Hydroxycinnamic acid can act as a starter unit for polyketide biosynthesis, leading to compounds of a mixed biosynthetic origin. Condensation with three acetate units forms a trihydroxychalcone (**1.62**) and thence a typical **flavanone**, narigenin (**1.63**). The glycoside of naringin is widespread and is one of the bitter-tasting substances in grapefruit juice (*Citrus paradisi*). The aglycones of the flavanoids are classified according to the oxidation state of the central pyran ring. Their extended conjugation leads to absorption in the visible region. Structural investigations into these pigments were begun by Willstätter in 1914 and continued by Karrer and Robinson in the 1920s.

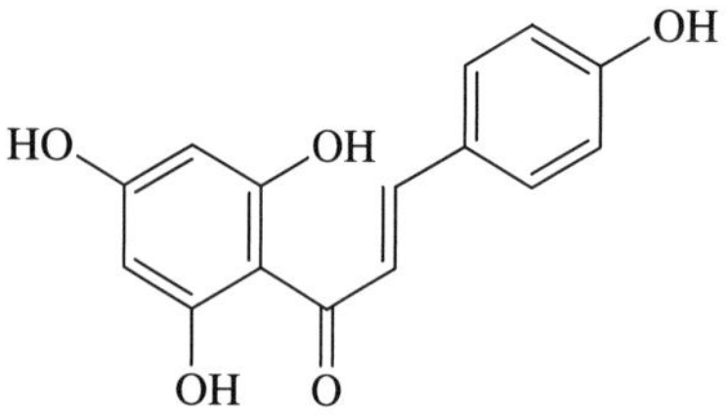

red

H$^+$

OH$^-$

blue

R = sugar

1.62 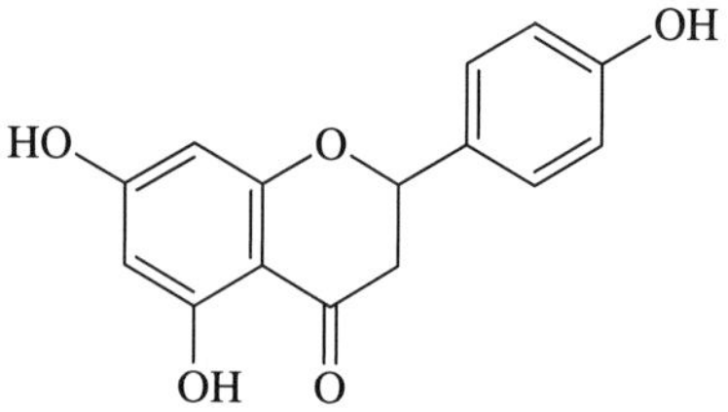   **1.63**

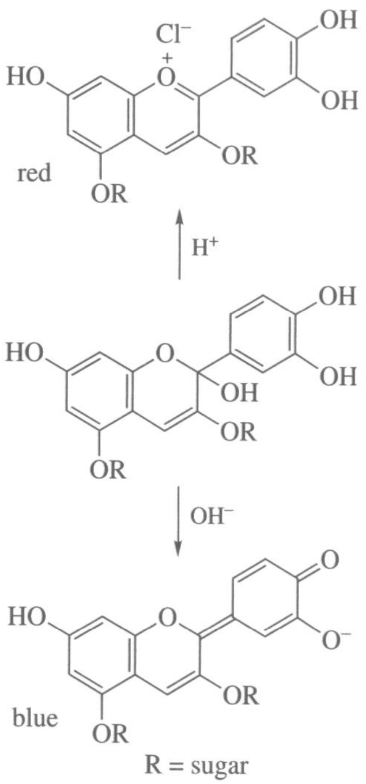

The **anthocyanins** are the glycosides of anthocyanidins such as cyanidin (**1.64**). The oxygen heterocyclic ring can form a salt. This gives rise to a different chromophore, and hence gives colours to flowers that range from red to blue, depending on the pH. In an acidic medium they form salts, but in a basic medium the phenols may ionize. The widespread

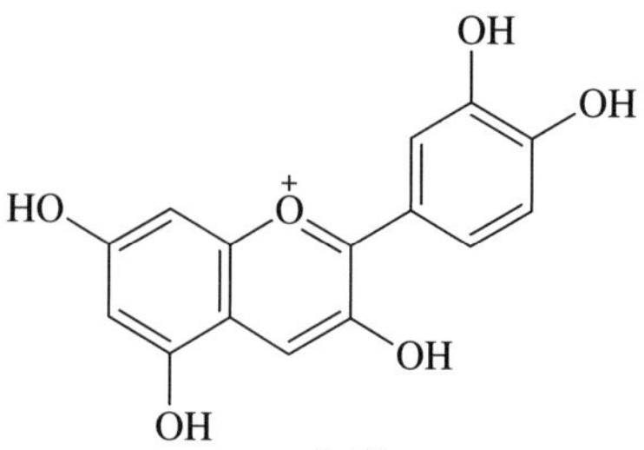

1.64 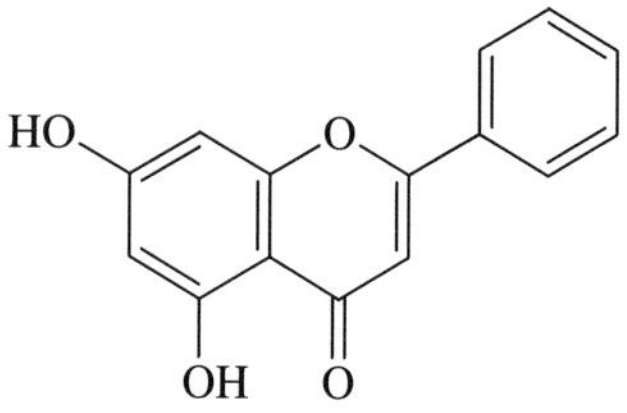   **1.65**

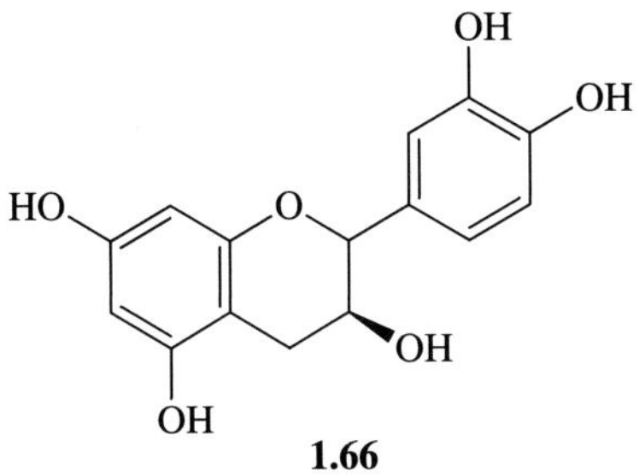

1.66 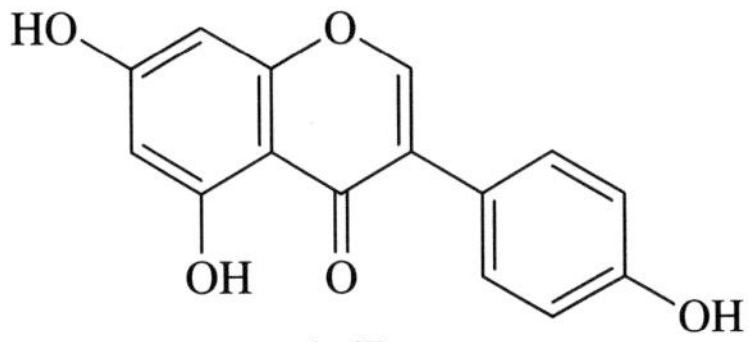   **1.67**

occurrence of this type of pigment has made them useful markers in plant classification (chemical taxonomy).

Chrysin (**1.65**), which gives the yellow colour to poplar buds, is an example of a **flavone**. Catechin (**1.66**) is colourless, but on oxidation it gives the brown colour which is associated with the discoloration of fruit. The **isoflavanoids** are formed from the flavanones by an oxidative rearrangement of the aromatic ring. They are found mainly in the legumes. One of them, genistein (**1.67**), is reported to have potential health benefits. **Stilbenes** (1,2-diphenylethenes) are formed biosynthetically by a different mode of cyclization of the condensation products between cinnamic acids and acetate units. These natural products include resveratrol, which contributes to the beneficial effects of red wine on heart disease.

Lignin is a polymeric material derived from the phenylpropanoid pathway, and forms part of the structural material of many woody plants.

The *Leguminosae* are an economically important family of flowering plants. The seeds are often borne in pods. Peas and beans are legumes.

Worked Problem 1.3

Q Identify the phenylpropanoid unit in the pigment isorhamnetin (**6**) which is found in wallflowers.

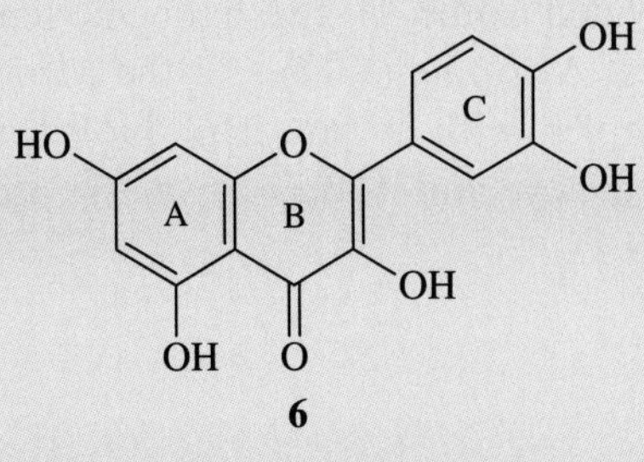

6 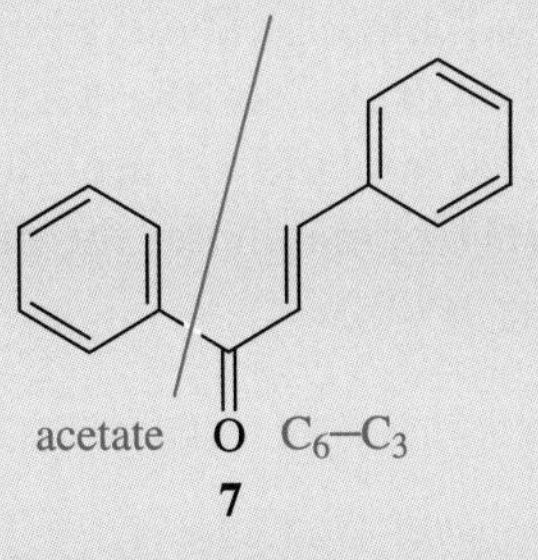   **7**

A Rings B and C comprise the phenylpropanoid unit and ring A is derived from acetate (see **7**). Note the different oxygenation patterns.

1.6   Alkaloids

Some of the first natural products to be isolated from medicinal plants were alkaloids. When they were first obtained from plant material in the early years of the 19th century, it was found that they were nitrogen-containing bases which formed salts with acids. Hence they were known as the vegetable alkalis or alkaloids. This ability to form salts and to complex with metal ions helped their separation and detection in the era before chromatography.

Many alkaloids have **neuroactive** properties and interact with the receptors at nerve endings. This is not surprising, since many alkaloids have fragments buried within their overall structure which resemble the natural substances (the neurotransmitters) that bind to these receptors.

Alkaloids may be grouped according to their plant sources, *e.g.* Aconitum, Amaryllidaceae, Cinchona, Curare, Ergot, Opium, Senecio and Vinca. Another classification is based on the structure of the ring system containing the nitrogen atom (*e.g.* piperidine, isoquinoline, indole). This can reflect their biosynthetic origin from amino acids such as ornithine, lysine, phenylalanine, tyrosine and tryptophan

A group of alkaloids possess **piperidine** or **pyrrolidine** rings. Thus the hemlock plant, *Conium maculatum*, produces the alkaloid coniine (**1.68**), which causes paralysis of nerve endings. The hemlock alkaloids are derived from polyketides and ammonia. The alkaloid piperine (**1.69**) and its *cis,cis* isomer chavicine are found in the fruit of the pepper *Piper nigrum*, and are responsible for the sharp taste of pepper. The fruit of the betel palm, *Areca catechu*, produces a mild stimulant, arecoline (**1.70**). The coca plant, *Erythroxylon coca*, is notorious for the production of cocaine (**1.71**), which has a paralysing effect on sensory nerve endings and produces a sense of euphoria. The roots, leaves and berries of a number of poisonous plants of the *Solanaceae*, including deadly nightshade (*Atropa belladona*), henbane (*Hyoscyamus niger*) and thorn apple (*Datura stramonium*), have been a rich source of therapeutically important tropane alkaloids. These plants provided some of the hallucinogenic "sorcerer's drugs" of the Middle Ages. Atropine (**1.72**) and the related epoxide, scopolamine, are two examples with a powerful biological activity. Atropine dilates the pupils of the eye and its derivatives are used in opthalmology.

Neuroactive = acts on the nervous system. Neurotransmitters are local hormones that are produced at nerve endings and transfer nerve impulses across the synaptic cleft.

piperidine

isoquinoline

indole

Cocaine blocks the re-uptake of the neurotransmitter dopamine in the brain and consequently increases its effect. The free alkaloid (crack cocaine) is absorbed more easily.

1.68

1.69

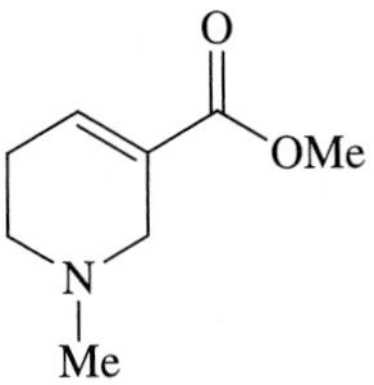

1.70 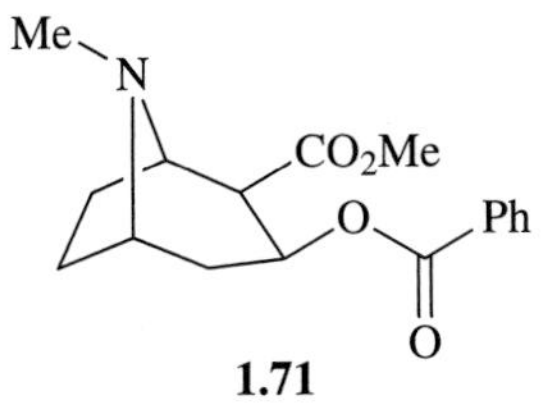 **1.71** 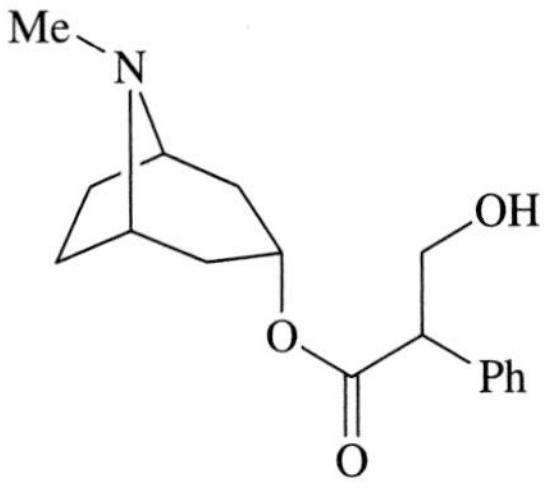 **1.72**

The tobacco plant, *Nicotiana tabacum*, produces the toxic alkaloid nicotine (**1.73**), which is the major neuroactive component of tobacco smoke. It is also used as an insecticide.

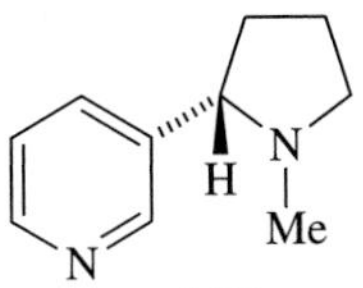

1.73 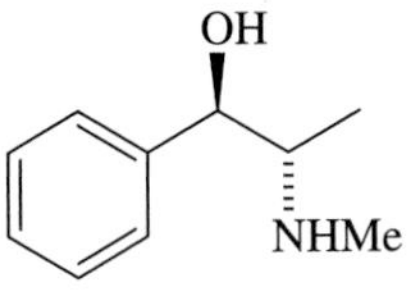 **1.74** 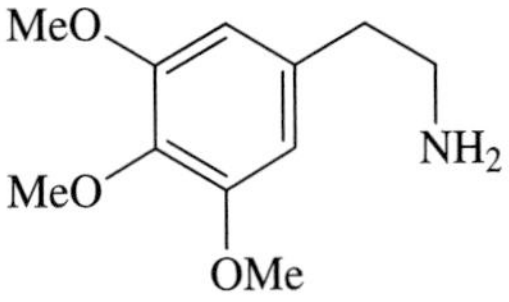 **1.75**

There are many alkaloids that are derived from simple phenylalkyl-amine C_6–C_3 units. The Chinese medicinal herb, *Ephedra sinica*, has been used for many years for the alleviation of bronchial problems. Extraction of the plant gave ephedrine (**1.74**) and its epimer, pseudoephedrine. Alkaloids such as mescaline (**1.75**) have been found in hallucinogenic plants such as the Mexican peyote cactus, *Lophophora williamsii* (Mescal buttons).

Ephedrine was at one time used as an anti-asthma drug, but it has been replaced by more effective synthetic analogues such as salbutamol. Ephedrine and pseudoephedrine are still used in some cold remedies.

The **benzylisoquinoline** alkaloids, which are formed via phenylalanine or tyrosine, are widespread. Like the anthocyanins, their occurrence has been used as a marker in the chemical taxonomy of plants. They have been found in various plants from the *Annonaceae, Lauraceae, Rhamnaceae, Ranunculaceae* and *Papaveraceae* families. The best-known source of these alkaloids is the **opium** poppy, *Papaver somniferum*, in which they co-occur with the more complex morphine series. Typical examples include papaverine (**1.76**), which has been used as a muscle relaxant. Berberine (**1.77**) is a yellow pigment from *Berberis* and *Mahonia* species which was used as a mild antibiotic in the treatment of sores.

Many of the tropical *Annonaceae* produce edible fruits. The *Rhamnaceae* include the buckthorns and the *Ranunculaceae* include the buttercup.

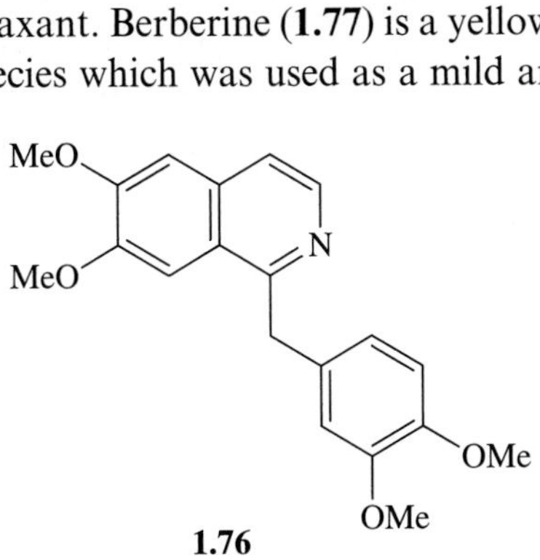

1.76 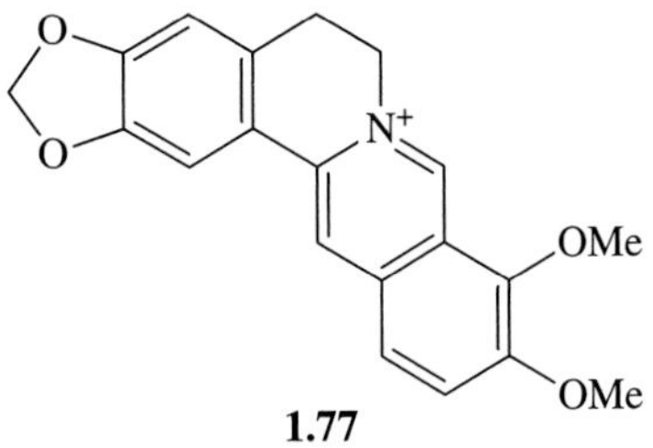 **1.77**

The pain-relieving and narcotic properties of opium were known in ancient times. In 1806, Sequin and, independently, Serturner obtained the crystalline alkaloid morphine (**1.78**) from opium. However, despite a great deal of work, it was not until 1925 when Gulland and Robinson were able to propose a structure for morphine. These alkaloids have been the basis of a very large amount of chemistry in an effort to separate the valuable pain-killing properties from their addictive narcotic properties.

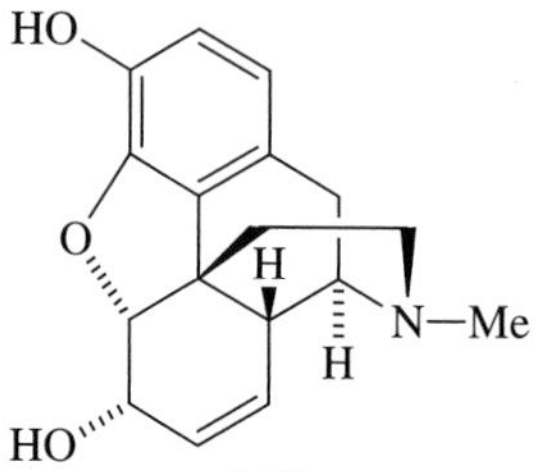

1.78 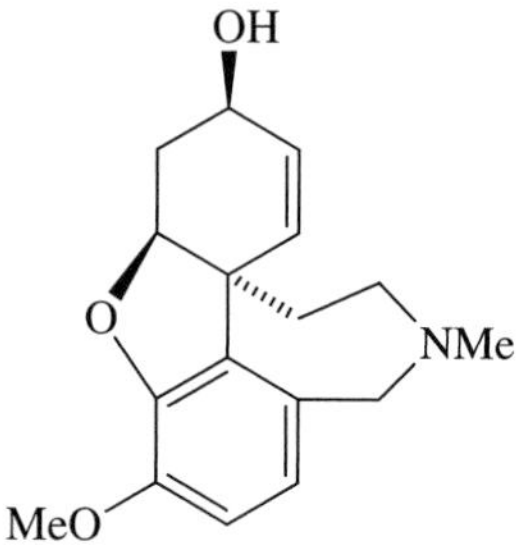   **1.79**

Bulbs of the *Amaryllidaceae* (daffodil family) contain alkaloids such as lycorine, which is widespread and quite toxic. Recently, galanthamine (**1.79**) has attracted interest because of its potential use in the treatment of Alzheimer's disease.

Malaria, which we now consider to be a disease of the tropics, used to be common in Europe. Jesuit missionaries returning to Europe in 1630 brought back cinchona bark, which cured the disease. In 1820, two French chemists, Caventou and Pelletier, were able to isolate the active principle quinine (**1.80**) from *Cinchona ledgeriana* bark. It contains the quinoline ring system and its correct structure was established by Rabe in 1907. It is used as an antimalarial drug and as a bitter substance in tonic waters.

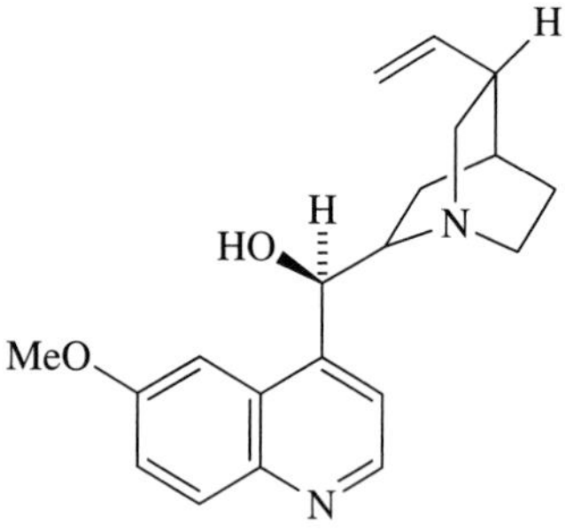

1.80 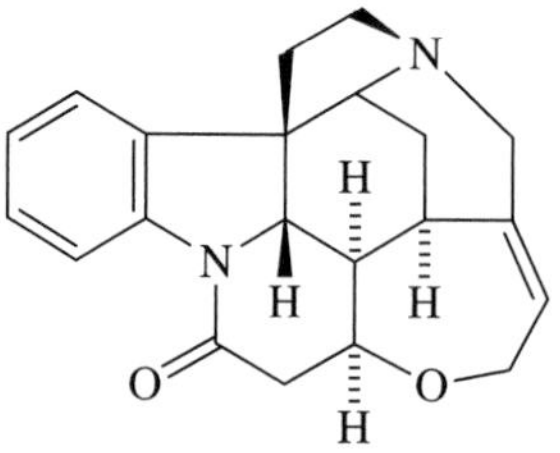   **1.81** 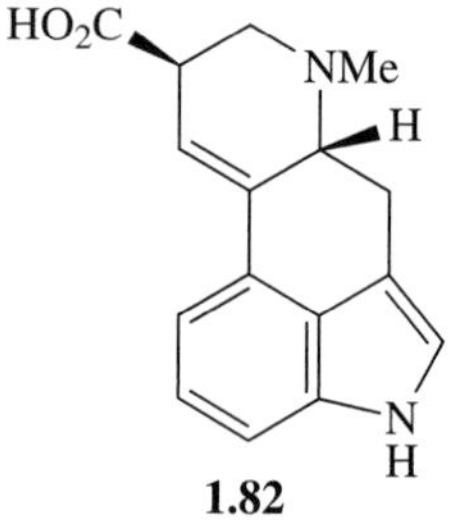   **1.82**

Another big family of alkaloids is the **indole** alkaloids. They are derived from the amino acid tryptophan and they occur in tropical plants from the *Apocyanaceae* (*Aspidosperma, Rauwolfia* and *Vinca*) and *Loganiaceae*

(*Strychnos*) families. These alkaloids include poisons such as strychnine (**1.81**), the structure of which was established by Robinson in 1946. Other compounds in this series have useful medicinal properties. Reserpine from *Rauwolfia* bark is used in the treatment of mental disease. The dimeric Vinca alkaloids are used in the treatment of leukaemia and other cancers. The ergot alkaloids, from the ergot fungus *Claviceps purpurea*, are also indole alkaloids and are amides of lysergic acid (**1.82**).

The purine carbon skeleton is found in the alkaloids caffeine (**1.83**) and theophylline, which are stimulants that occur in coffee and tea. Kinetin (**1.84**) is a cytokinin plant growth substance that stimulates cell division in plants.

Although most of the *Apocyanaceae* are tropical plants, the common periwinkle, *Vinca minor*, is one of the few members of this family which grow in temperate climates.

Strychnos nux vomica was used as a rodent poison and features in various "murder mysteries".

The metabolites of the ergot fungus, *Claviceps purpurea*, growing on rye caused ergotism, known in the Middle Ages as St. Anthony's Fire, a severe and often fatal disease.

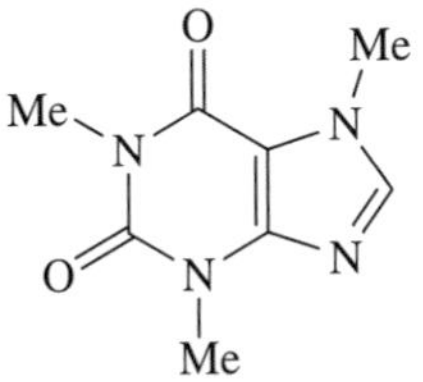

1.83 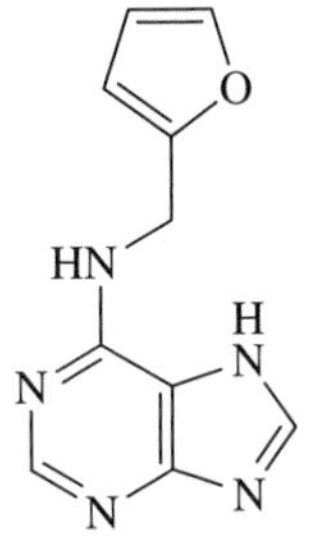   **1.84**

In some alkaloids, such as the steroidal alkaloids and the toxic diterpenoid alkaloids from *Aconitum* species, the nitrogen atom is inserted after the carbon skeleton has been formed. In a biosynthetic sense these might be considered as terpenoids.

The roots of various *Aconitum* species such as A. *napellus* (monkshood) have had various uses as folk medicines and as poisons. Some species are used as ornamental plants.

Worked Problem 1.4

Q   The metabolite clavicipitic acid (**8**) is produced by the ergot fungus, *Claviceps purpurea*. Identify the building blocks.

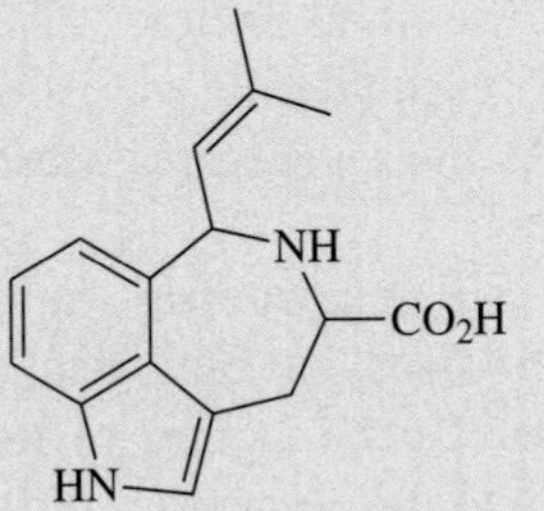

8   **9**

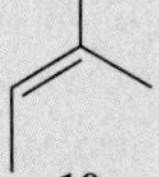

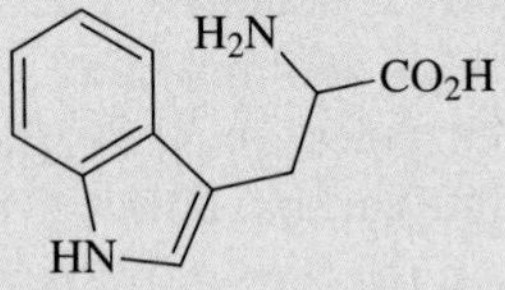

A   The alkaloid is derived from the amino acid tryptophan (**9**) and a dimethylallyl terpenoid unit (**10**).

1.7   Antibiotics Derived from Amino Acids

The alkaloids are not the only group of secondary metabolites that are derived from amino acids. Amino acids not only form the building blocks for the large peptides and proteins but also for smaller peptides that are converted into the β-lactam antibiotics such as the penicillins (**1.85**) and cephalosporins (**1.86**) (see Chapter 4). The diketopiperazine antifungal agents produced by *Trichoderma* and *Gliocladium* species, such as gliotoxin (**1.87**), are also derived from amino acids.

Trichoderma and *Gliocladium* species are common soil fungi.

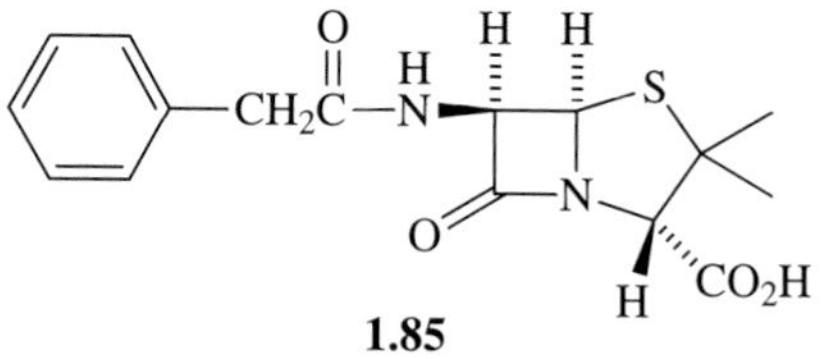

1.85

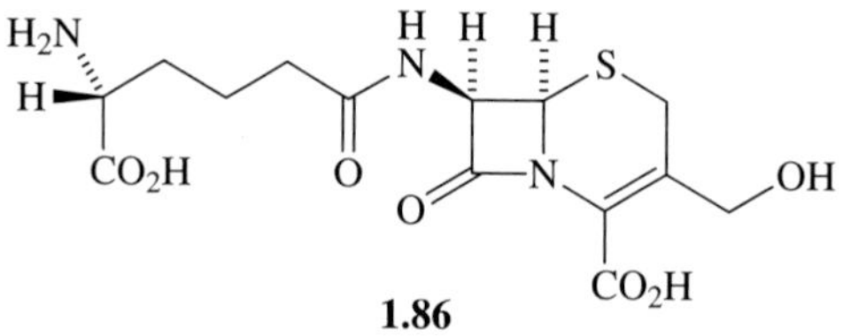

1.86 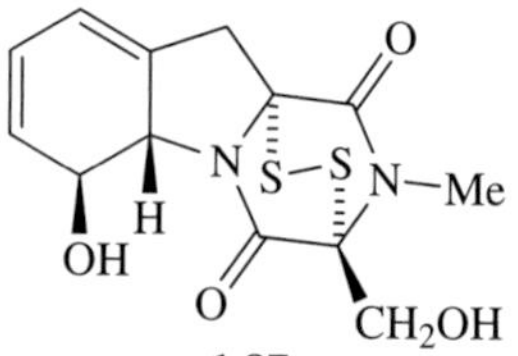   **1.87**

1.8   Vitamins

There are a number of natural products that are essential for life but which cannot be produced by the body. The recognition by Hopkins, Grijns and others at the start of the 20th century that diseases such as scurvy, rickets and beri-beri, which arose from deficiencies in the diet, might be associated with a requirement for particular compounds present in foods, led to the search for these essential factors. These natural products became known as vitamins. As each vitamin was isolated, it was identified by a letter of the alphabet. As their structures became known, these compounds acquired a trivial name.

A co-enzyme is a non-protein part of an enzyme that participates in the catalytic function of the enzyme.

A number of vitamins play a role as co-enzymes in the function of particular enzymes such as those involved in biological oxidation, reduction and carboxylation. This biological function of the vitamins places them in the class of primary metabolites. However, the elucidation of their structures and aspects of their chemistry and biosynthesis link them with secondary metabolites.

In the period following 1920, bioassay-guided fractionation using test animals led to the isolation of various vitamins. They are present in their natural sources in low concentrations, and consequently very small amounts of material were available for structural studies. Nevertheless, the structures of many of the vitamins were established in the 1930s and confirmed by synthesis. Material then became available for biological studies and dietary supplements.

The presence of the fat-soluble vitamin A in egg and in fish oils was established in 1913–1915, and its structure was elucidated by Karrer in 1931. The recognition that the disease beri-beri, which was prevalent in South East Asia, was a deficiency disease and that it could be cured by an extract obtained from rice polishings, led Funk in 1911 to coin the name vitamin for these essential dietary factors. Vitamin B_1 (thiamine) was isolated in 1926 and in quantity by Williams in 1934. Its structure (**1.88**) was established in 1936 and confirmed by synthesis. Other members of this group of vitamins include riboflavin (vitamin B_2, **1.89**), pyridoxamine (vitamin B_6, **1.90**), biotin (**1.91**), folic acid (**1.92**) and vitamin B_{12} (**1.93**). Vitamin B_{12} was isolated in 1948 and its structure established by a combination of chemical work (Todd) and X-ray crystallography (Hodgkin) in 1954. It is one of the most potent of the vitamins, being active in the treatment of pernicious anaemia at the level of micrograms. Vitamin C (ascorbic acid, **1.94**) (see Chapter 4) is important in the prevention of scurvy. Its structure was established in 1932. Vitamin D, a steroid, has already been mentioned. Vitamin E, tocopherol (**1.95**), is an antioxidant and vitamin K, a quinone (**1.96**), is an antihaemorrhagic factor.

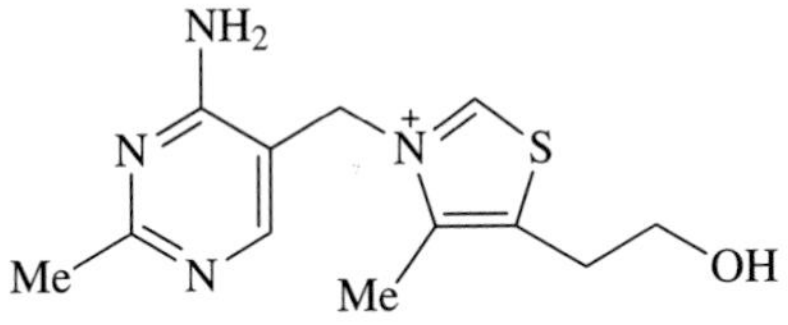

1.88

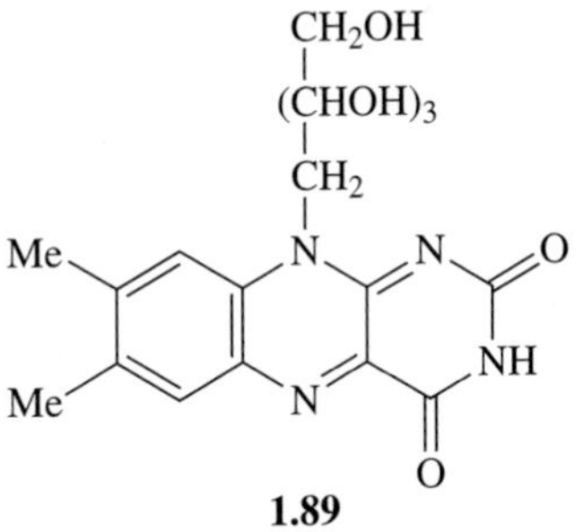

1.89

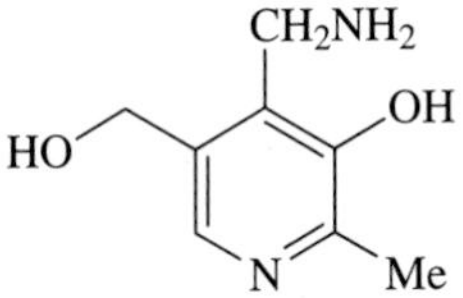

1.90

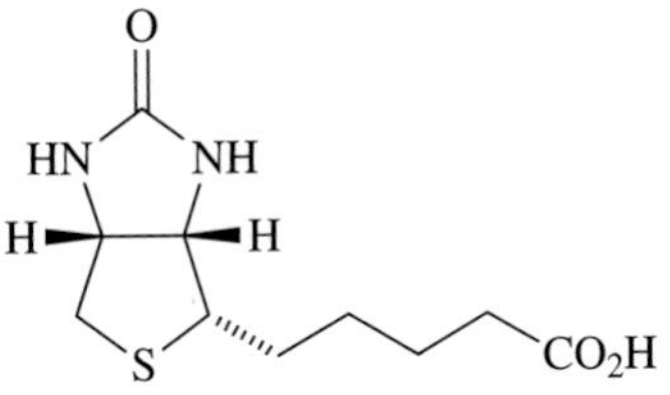

1.91

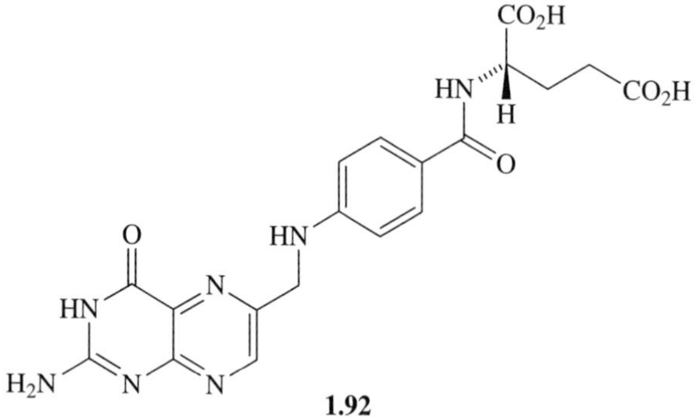

1.92

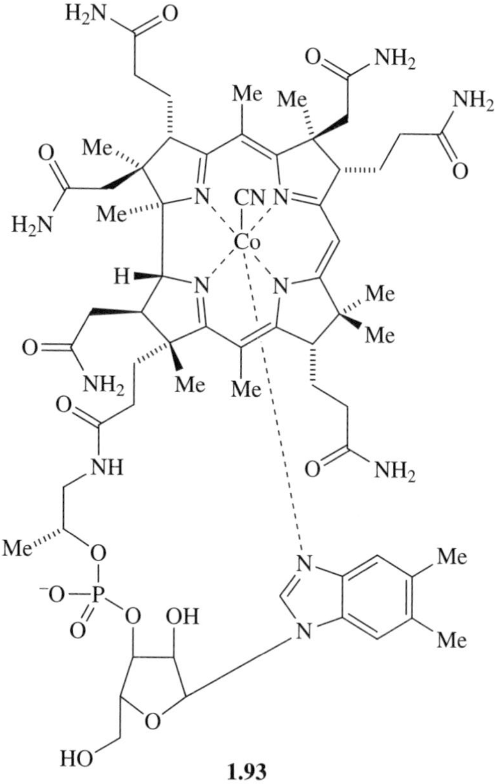

1.93

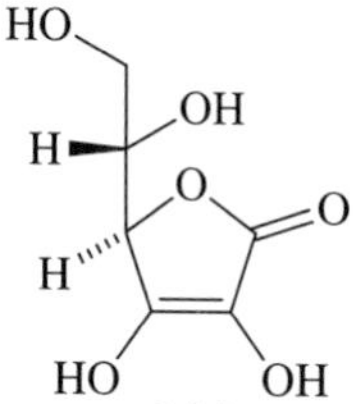

1.94

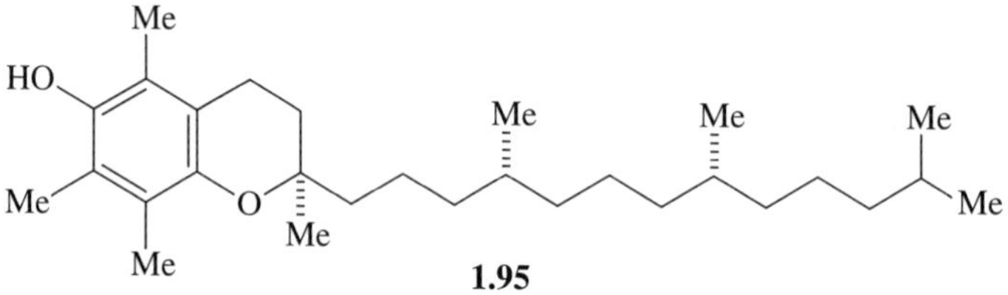

1.95

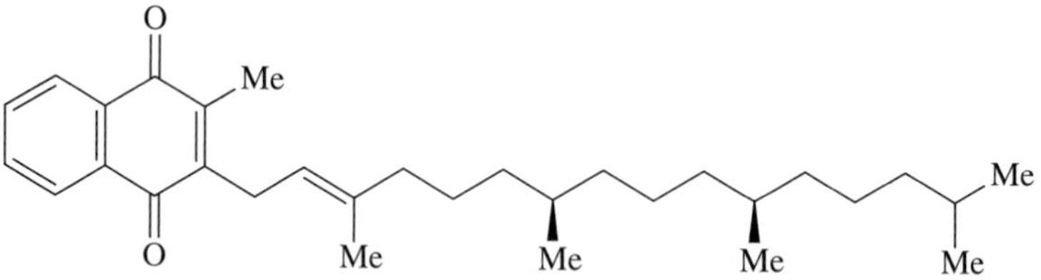

1.96

1.9 Chemical Ecology

Ecology is the study of the interaction between an organism and its environment.

Many natural products were originally investigated because of their medicinal, perfumery or culinary value to man. However, during the latter part of the 20th century, increasing attention has been paid to the biological function of natural products, and to their ecological role in regulating interactions between organisms. Developments in instrumental methods have meant that very small amounts of material can be detected and identified and their effects, particularly on insects, can be observed.

Some plants exert their dominance of an area by the production of compounds which inhibit the germination of seeds and prevent the growth of seedlings. This harmful effect of one plant on another, caused by the production of a secondary metabolite, is known as **allelopathy**. One of the best-known examples is the production of the quinone juglone (**1.97**) by the walnut tree, *Juglans nigra*. This compound inhibits seed germination and is toxic to other plants. Some monoterpenes, such as 1,8-cineole (**1.98**) and camphor (**1.23**), which are produced by the sagebrush (*Salvia leucophylla*), act as seed germination inhibitors. The presence of these monoterpenes allows this plant to dominate the vegetation in some of the drier parts of California until it is destroyed by bush fires. The monoterpenes produced by *Eucalyptus* species have a similar effect. The alkaloid hordenine (**1.99**), which is produced by the roots of barley, *Hordeum vulgare*, has an allelopathic effect.

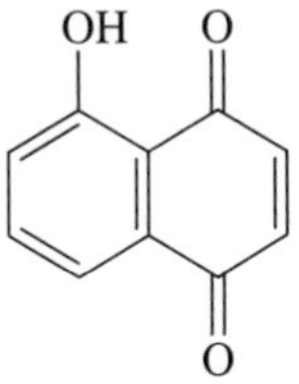

1.97	**1.98**	**1.99**

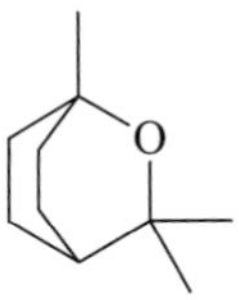

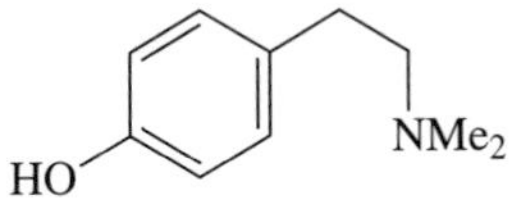

Fungi attack plants by producing **phytotoxins** together with enzyme systems that digest plant tissues. In response to this attack the plant produces **phytoalexins**, which act as natural antifungal agents. Some examples are rishitin (**1.100**), which is produced by the potato, *Solanum tuberosum*, and phaseolin (**1.101**), which is produced by the bean, *Phaseolus vulgaris*. Many of the phenols produced by plants give protection against microbial attack.

A phytoalexin is a natural product which is produced by a plant in response to stress.

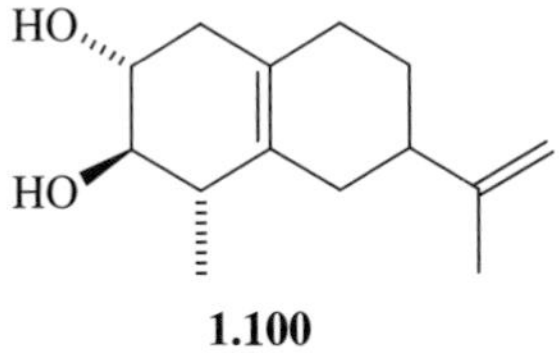

1.100	**1.101**

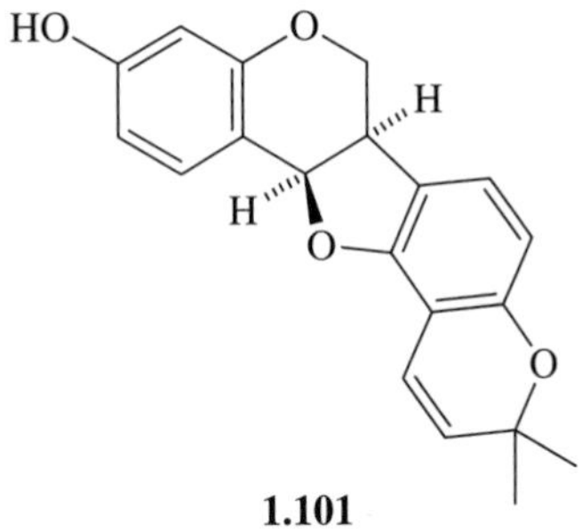

Fungi competitively colonize the soil. *Trichoderma* species are particularly invasive. Because they are not serious plant pathogens, they have been used as bio-control agents to inhibit the development of other organisms that are plant pathogens. They achieve this dominance by producing volatile antifungal agents, such as pentylpyrone (**1.102**), which permeate the soil around the developing organism. They then produce a second "close contact" antifungal agent (*e.g.* gliotoxin, **1.87**), which enables the *Trichoderma* species to compete effectively with other organisms.

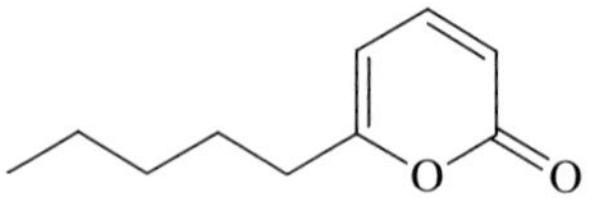

1.102 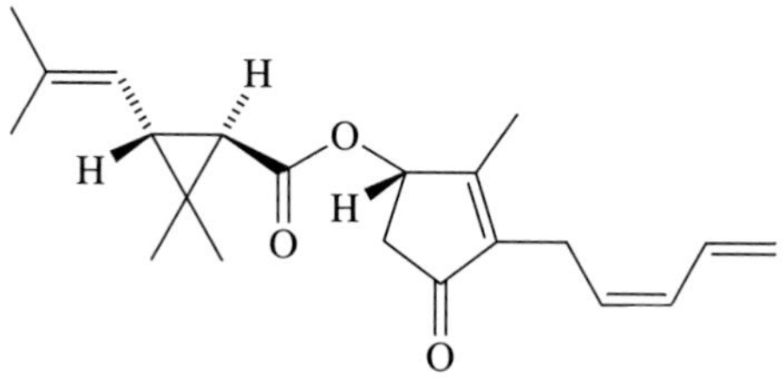   1.103

The synthetic insecticides, such as permethrin, are based on the pyrethrin structure.

Many plant–insect relationships are determined by the presence of secondary metabolites. These compounds may be either deterrents or attractants. Plants have developed antifeedants and insecticides as defensive agents against insect attack. Some of these compounds, such as nicotine (**1.73**), the pyrethrins (*e.g.* **1.103**) from *Chrysanthemum cinearifolium* and rotenone from Derris root, have formed the basis of commercial insecticides.

There are some interesting relationships involving the toxic cardenolides (see p. 13) of the *Asclepediaceae* (*e.g.* caliotropin) produced by the milkweed, *Asclepias curassavica*. These steroidal glycosides are toxic to many mammals and insects and can affect the heart. However, Monarch butterflies sequester and accumulate these compounds. The glycosides are toxic to birds that are predators of the butterflies and so they act as protective agents. These butterflies also take up the *Senecio* pyrrolizidine alkaloids (*e.g.* retronecine, **1.104**) as protective agents. The toxic alkaloids produced by plants can act as feeding deterrents for animals. The *Senecio* alkaloids are poisonous to horses and cattle and ragwort (*Senecio jacobaea*) has achieved notoriety in this connection.

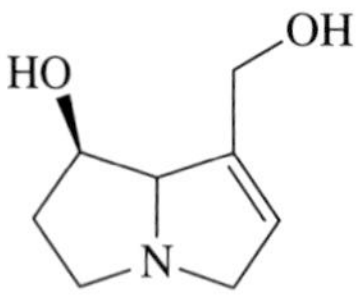

1.104 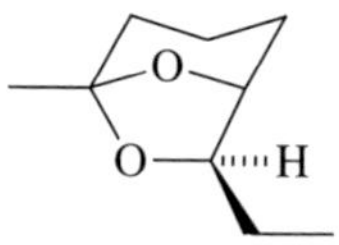   1.105 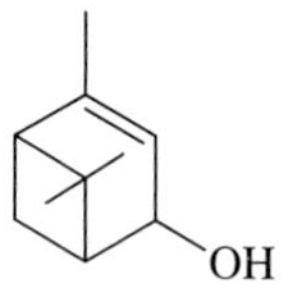   1.106

There are examples of insects "borrowing" secondary metabolites from plants and then modifying them to make insect **trail substances** for feeding and mating purposes. For example, the bark beetle, *Dendroctanus brevicomis*, is attracted to a pine tree, *Pinus ponderosa*, by its volatile terpenes. The female produces *exo*-brevicomin (**1.105**) and the related frontalin as attractants, but when the population has reached a particular level the beetles begin to modify the α-pinene produced by the plant to verbenol (**1.106**), which acts as a deterrent.

Insects may be attracted to particular plants by specific secondary metabolites in order to lay their eggs. For example, allyl isothiocyanate (**1.107**), produced from sinigrin in the cabbage, acts as an **attractant** for the cabbage white butterfly, *Pieris brassicae*. The volatile monoterpenes produced by flowers act as attractants for bees, butterflies and moths for pollination. The components of the floral fragrance often have a synergistic action as attractants. Insects that are involved in the pollination of a particular species respond to a specific combination of monoterpenes and to the colour of the flower.

$$CH_2{=}CHCH_2N{=}C{=}S$$
1.107

1.108   **1.109**   **1.110**

Many aspects of insect behaviour are regulated by chemical stimuli. The term **semiochemical** is used to describe these signalling substances. **Pheromones** are semiochemicals involved in intra-species communication. The sex pheromones of moths have received a great deal of attention. Many of the compounds, although species specific, are simple esters derived from fatty acids or long-chain alcohols, or are derived from monoterpenes such as eldanolide (**1.108**). The aggregation pheromones of bark beetles such as ipsdienol (**1.109**) have been studied because of the economic damage caused by these insects. Trail pheromones play an important role in communication between social insects such as ants. The leaf-cutting ant (*Atta texana*) uses a simple pyrrole for this purpose. Insects can also produce substances that signal alarm: β-farnesene (**1.110**) is an aphid alarm pheromone.

A number of natural products are involved in the hormonal regulation of insect development. Many insects undergo metamorphosis through various juvenile forms before attaining the adult stage. The **insect juvenile hormones** (*e.g.* **1.111**) that maintain this status are related to the sesquiterpenes, but the hormone ecdysone (**1.112**), which regulates the development of the adult insect from a cocoon, is a steroid. Curiously, plants also produce these ecdysteroids as protection against excessive insect populations.

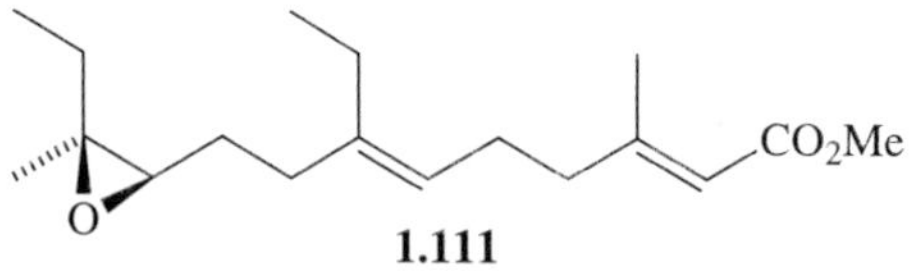

1.111 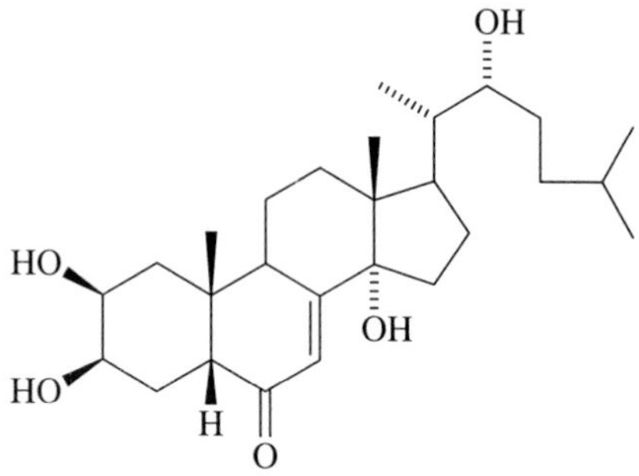   **1.112**

In this survey of the biological role of natural products it should be apparent that any one biological activity is not necessarily confined to one specific group of natural products.

1.10 The Isolation of a Natural Product

Secondary metabolites, with some exceptions, occur in amounts that are less than 0.01% of the dry weight of the plant. Extraction of 1 kg of dry plant material is likely to yield less than 100 mg of a natural product. These compounds may be unstable and present as part of a complex mixture. The isolation, separation and purification of these natural products require considerable skill.

The source of a secondary metabolite requires proper identification and a voucher specimen needs to be retained. Within the same species there are sometimes chemotypes, each with a particular composition. Some compounds are found in the roots, some are components of the bark, and others may be found in the leaves, the flowers or the fruit. Some compounds play a seasonal role in the plant, for example as insect antifeedants. Thus the part of the plant and the place and date on which the plant was collected should all be recorded.

Micro-organisms are usually deposited in national culture collections. The production of microbial metabolites often depends upon the medium on which the micro-organism is grown and on other fermentation details. Some fungal metabolites are retained in the fungal mycelium, whilst others are excreted into the broth. Insects, marine organisms and fungi that were collected in the wild may have stored and modified compounds which they had obtained from their food.

Natural products may be obtained from the crushed biological material by **extraction** with a solvent such as petroleum ether, chloroform

Chemotype = a sub-species producing a particular natural product which is not necessarily found in all examples of the species.

(trichloromethane), ethyl acetate (ethyl ethanoate) or methanol. Several solvents of increasing polarity may be used. Thus lipid material (waxes, fatty acids, sterols, carotenoids and simple terpenoids) can be extracted with non-polar solvents such as petroleum ether, but more polar substances such as the alkaloids and glycosides are extracted with methanol, aqueous methanol or even hot water. Many alkaloids are present as their salts with naturally occurring acids such as tartaric acid.

Commercial extractions may use tonne quantities of plant material, and a range of different extraction procedures including steam distillation have been used. Recently, commercial procedures have been developed using super-critical carbon dioxide as a mild solvent, but because of the pressures involved this requires quite complicated apparatus.

The initial extraction is then followed by a **separation** into acidic, basic and neutral fractions. A typical fractionation is set out in Scheme 1.1. A solution of the extract in an organic solvent (such as ethyl acetate) is shaken with an inorganic base (such as aqueous sodium hydrogen carbonate) to remove the carboxylic acids as their water-soluble sodium salts. The more weakly acidic phenols may only be extracted with a sodium hydroxide solution. Extraction of the original solution with an acid such as dilute hydrochloric acid will remove the bases such as the alkaloids as their salts. The neutral compounds remain behind in the organic phase. The acids and the phenols may be recovered from the aqueous solution of their sodium salts by treatment with dilute hydrochloric acid and re-extraction with an organic solvent, and the bases may be recovered by treatment of their salts with ammonia and re-extraction.

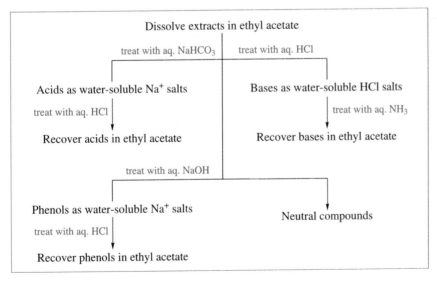

Scheme 1.1 Separation of an extract into acidic, phenolic, basic and neutral fractions

Although some abundant natural products may be obtained merely by extraction, a simple fractionation or partition and crystallization, the majority are obtained after further careful chromatography. A typical example might involve chromatography on silica or alumina and elution with increasing concentrations of ethyl acetate in petroleum ether. The chromatographic separation may be monitored by a bioassay or by thin layer chromatography (TLC). A number of useful spray reagents have been developed which produce coloured TLC spots indicative of particular classes of compound.

In the course of a separation, reactions such as ester hydrolysis, auto-xidation and rearrangement may occur, leading to the formation of artefacts. Hence efforts are made to ensure that a separation is chemically mild.

Natural products often co-occur in closely related series, for example as the mono-, di- and trihydroxy derivatives of the same parent compound. Since this relationship may facilitate structure determination, it is helpful to characterize and examine not only the major product from an isolation but also the minor components.

An artefact, in this context, is a compound arising from a natural product by human intervention.

1.11 The Stages in Structure Elucidation

In examining the strategies for natural product structure determina-tion, it is possible to discern four overlapping stages. First there is the preliminary **characterization** in which the physical constants are determined, the molecular formula is established and the functional groups are identified. At this stage it is often possible to recognize the class of natural product to which the compound belongs. It may be apparent that the compound is known from another source or is just a simple derivative of another compound of known structure. A simple chemical interrelationship may then be enough to establish the structure of the unknown compound.

The second stage is one of **structural simplification** in order to identify the underlying **carbon skeleton**. Chemically, this work involves the selective removal of the various functional groups and the dissection of the carbon skeleton into identifiable fragments. Spectroscopically, it involves identifying adjacent groups of atoms by, for example, their NMR characteristics. Evidence may accumulate concerning the position of the functional groups on the carbon skeleton. The determination of the position and relative **stereochemistry** of the functional groups forms the third phase. There is an interesting contrast in the use of spectroscopic and chemical methods between the first and third stages of this strategy. In the first stage, the methods are used to obtain information about the separate functional groups, but in the third phase the methods are used to reveal interactions between groups. The fourth stage involves establishing the **absolute stereochemistry** of the molecule. Although this is sometimes assumed, it is important to confirm it in order to understand the biological activity of a natural product. A few natural products are found in both enantiomeric forms.

When a natural product or a simple derivative crystallizes particularly well, X-ray crystallography can provide an unambiguous way of establishing the structure. Using a modern diffractometer, structures can be obtained in a matter of hours rather than days. However, unless a heavy atom such as bromine is present or a derivative is made with an optically active compound of known absolute stereochemistry, the structure determined this way will contain only relative stereochemical information.

When a structure has been proposed, it is helpful to rationalize this in biogenetic terms. An unambiguous partial synthesis from a compound of known structure, or a total synthesis, will finally provide the ultimate proof of structure.

Summary of Key Points

1. Primary metabolites are natural products that are found in all cells, but secondary metabolites are natural products that are restricted in their occurrence.

2. Secondary metabolites have been studied because of their biological activity, or for chemosystematic or ecological reasons.

3. Secondary metabolites may be classified as polyketides, terpenoids and steroids, phenylpropanoid (C_6–C_3) compounds, alkaloids and carbohydrates, based on their biosynthetic building blocks.

4. Acetyl co-enzyme A, isopentenyl pyrophosphate and shikimic acid are the building blocks for the polyketides, terpenoids and phenylpropanoid compounds, respectively.

5. Alkaloids may be classified in terms of the nitrogen-containing ring system which may reflect the amino acids (lysine, ornithine, tyrosine and tryptophan) which are involved in their biosynthesis.

6. Natural products are obtained from plant sources in quantities of 10–100 mg/kg dry weight by solvent extraction, partition into acidic, basic and neutral fractions, and chromatography.

7. The elucidation of the structure of a secondary metabolite involves characterization, determination of the carbon skeleton, establishing the position of the functional groups, and determining the relative and absolute stereochemistry.

Problems

1.1. To which class of natural product does each of the following belong?

(a)

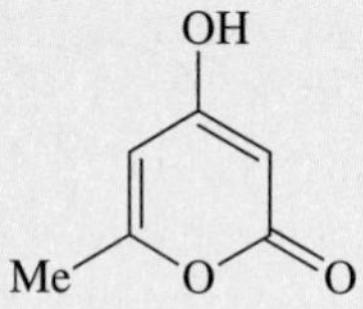

(b)

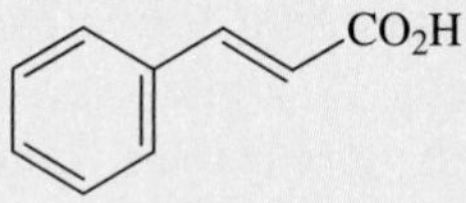

(c)

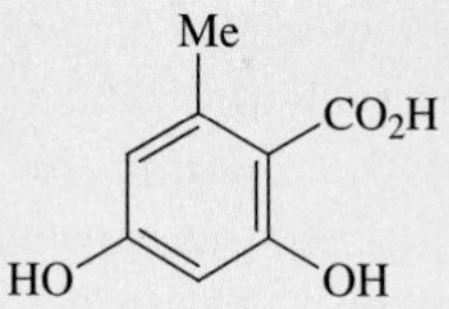

(d)

(e)

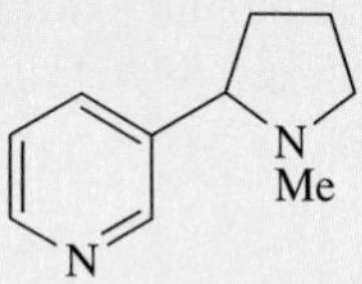

(f)

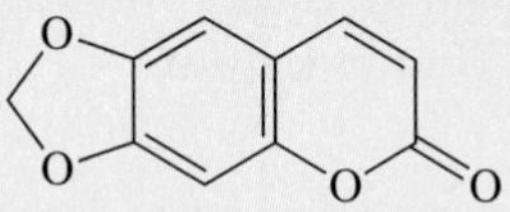

(g)

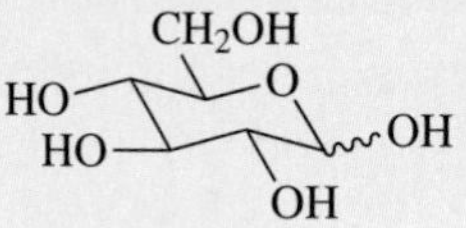

(h)

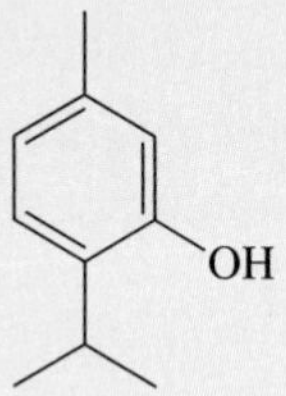

(i)

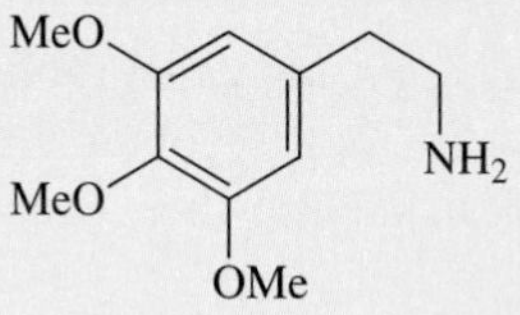

1.2. Indicate, using bold lines, the constituent isoprene units in the following terpenoids:

(a)

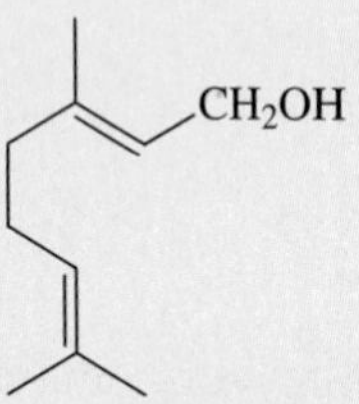

(b)

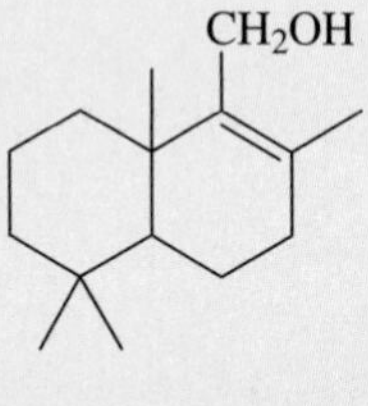

(c)

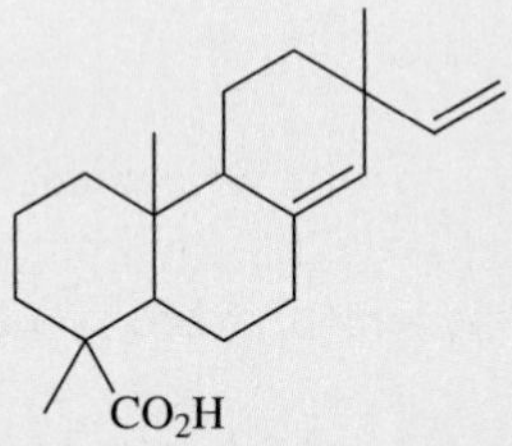

1.3. Indicate, using bold lines, the constituent acetate units in the following:

(a)

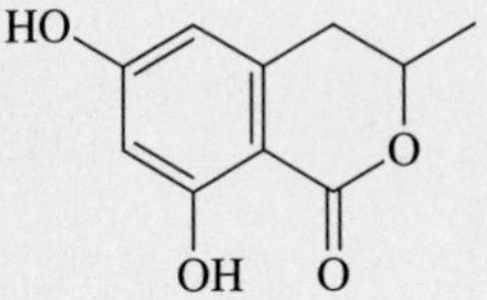

(b)

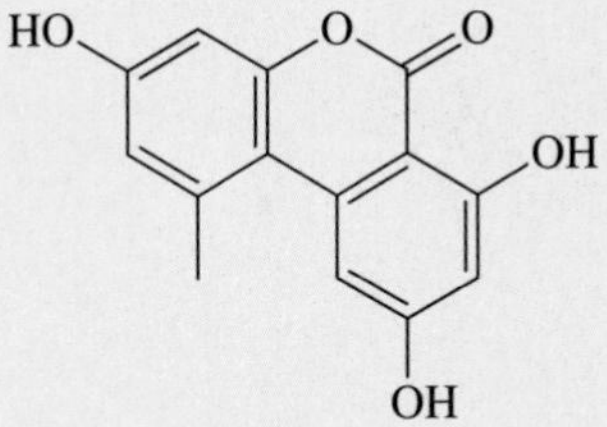

(c)

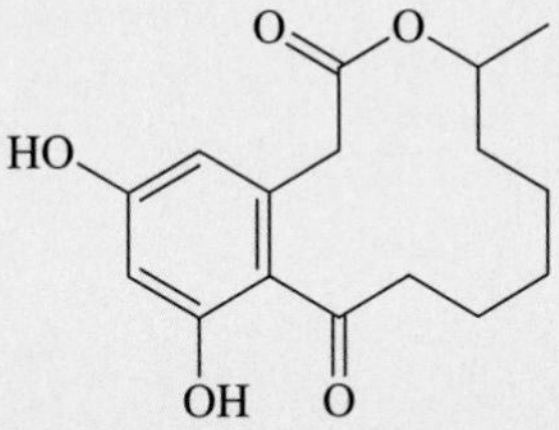

1.4. Show, by means of a flow chart, how you would isolate the alkaloids from a sample of plant material.

1.5. The fungal metabolite muscarine (**A**) binds to certain receptors for the neurotransmitter acetylcholine (**B**). Indicate, on their structures below, the structural similarity between muscarine and acetylcholine.

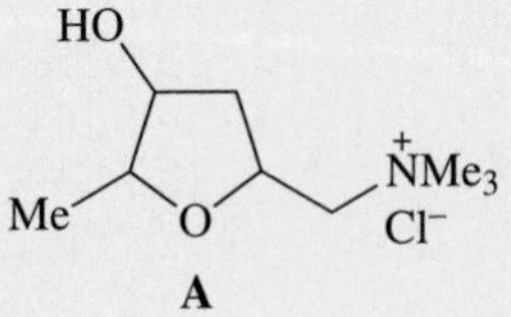

A

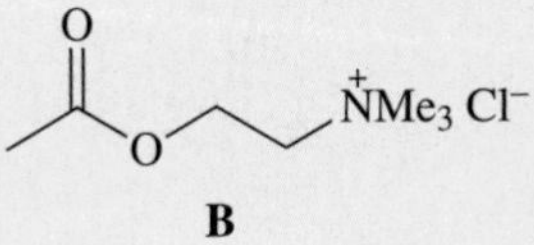

B

2

The Characterization and Determination of the Carbon Skeleton of a Natural Product

Aims

The aim of this chapter is to describe the information that is used to characterize a natural product and to establish its underlying carbon skeleton. By the end of this chapter you should understand:

- The role of physical, analytical, chromatographic and spectroscopic criteria in characterizing a natural product
- The role of structural simplification in the determination of the carbon skeleton
- The part played by oxidative and reductive degradation, dehydrogenation, elimination reactions and alkaline degradation
- The role of spectroscopic methods, particularly NMR spectroscopy, in determining the carbon skeleton

2.1 The Characterization of a Natural Product

The first stage in the structure determination is to establish the purity and to characterize the compound in terms of its elemental composition, empirical and molecular formulae. Some of the early literature on natural products is confused because the materials being studied were mixtures. The **criteria of purity** and the basic **analytical data** that are used fall into four groups:

- **Physical criteria**: melting point or boiling point, optical rotation, refractive index.

- **Analytical criteria**: elemental composition determined by combustion analysis or high-resolution mass spectrometry, relative molecular mass.
- **Chromatographic criteria**: single spot on thin layer chromatography, or single peak on gas chromatography or high-pressure liquid chromatography determined in several systems.
- **Spectroscopic criteria**: consistent relative integrals in the ^{1}H NMR spectrum and consistent absorption in the infrared and ultraviolet spectrum.

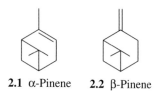

2.1 α-Pinene   **2.2** β-Pinene

Many secondary metabolites occur as mixtures of closely related compounds. For example, α-pinene (**2.1**) from oil of turpentine often contains some of its double bond isomer, β-pinene (**2.2**). The criteria that are used to assess purity must be adequate to detect the presence of close relatives.

The micro-analytical determination of the percentage elemental composition gives the empirical formula, which, when taken together with the relative molecular mass, leads to the molecular formula. Present-day methods of combustion analysis use 2–3 mg of material, with the amount of the gases such as carbon dioxide and water that are formed being determined by gas chromatography. The molecular formula of a compound can be used to establish the number of double bonds or rings. The number of double bond equivalents may be calculated from the molecular formula as follows. Every time a double bond or a ring is introduced into an initially saturated acyclic hydrocarbon which contains C_nH_{2n+2} atoms, two hydrogen atoms are removed. Thus, a monoalkene contains C_nH_{2n} atoms, a diene contains C_nH_{2n-2} atoms, and so forth. A saturated monocyclic hydrocarbon may be represented by the general formula C_nH_{2n}. Hexane is C_6H_{14}, hexatriene is C_6H_8, cyclohexane is C_6H_{12} and cyclohexatriene (benzene) is C_6H_6. In order to calculate the number of double bond equivalents, the actual number of hydrogens is subtracted from the theoretical maximum for the corresponding open-chain hydrocarbon with the same number of carbon atoms. This difference is then divided by two. If the molecule contains oxygen, it is disregarded for the purpose of the calculation. In this context, a carbonyl group counts as a double bond equivalent. If the molecule contains a halogen, it is replaced by hydrogen, and if it contains nitrogen, it is replaced by CH. Thus pyridine (C_5H_5N) is equivalent to benzene (C_6H_6) for the purpose of calculating the number of double bond equivalents.

Worked Problem 2.1

Q   Carotol, $C_{15}H_{26}O$, is an alcohol which has been obtained from carrot seed oil. On catalytic hydrogenation it gives a saturated alcohol, $C_{15}H_{28}O$. How many rings are present in carotol?

A   A C_{15} saturated acyclic compound is $C_{15}H_{32}$ (C_nH_{2n+2}). There is a discrepancy of six hydrogens with carotol and hence there are three double bond equivalents. One of these is an alkene and since carotol is an alcohol (rather than a ketone) there are two rings.

2.2   Spectroscopic Characterization

The first objective of the spectroscopic characterization of a secondary metabolite is to identify the functional groups that are present and possibly some significant molecular fragments. This stage will determine the course of the subsequent detailed investigation. The initial inspection of the spectroscopic data may be followed later by a more intensive study, often using specialized techniques as the structure is probed more deeply. Although the information from each of the spectroscopic methods may be discussed separately, the structural conclusions drawn from the various methods must be mutually consistent. Thus the identification of a conjugated system from the UV spectrum must be sustained by the identification of the corresponding sp^2 carbon resonances in the ^{13}C NMR spectrum. The identification of a carbonyl group in the IR spectrum also has implications in the ^{13}C NMR spectrum.

It is worth remembering that the majority of spectroscopic correlations were obtained empirically by examining the spectra of compounds, including many natural products of known structure. For a detailed discussion of spectroscopic methods in organic chemistry, the reader is referred to one of the standard textbooks on the topic. The following is intended only to describe the salient features that affect natural product structure determination.

2.2.1   Mass Spectrometry

The **low-resolution mass spectrum** may reveal the **molecular ion** and hence the relative molecular mass of the compound. For some natural products, ionization by **electron impact** (EI) is too vigorous a process and considerable fragmentation occurs. Consequently, the molecular ion is of relatively low abundance and may not be detectable. Soft ionization techniques, such as **chemical ionization** (CI), may be more rewarding, although the ion of highest mass that is obtained may be a composite of

A molecular ion is an ion produced from the intact molecule which has not undergone fragmentation.

Chemical ionization is when the substance is ionized by interaction with ions produced from a carrier gas such as ammonia or methane.

Field desorption is when ions are produced from the substance coated on a wire by local very high electrical potentials.

Fast atom bombardment is when the substance, usually dispersed in glycerol, is ionized by impact with high-energy xenon atoms.

the molecular ion and the ionizing gas (*e.g.* $M + NH_4^+$). Other techniques such as **field desorption** and **fast atom bombardment** (FAB), which are particularly applicable to relatively non-volatile polyoxygenated compounds, have extended the mass range available to mass spectrometric analysis into that of polypeptides and glycosides.

The elemental composition may be apparent from the molecular ion and confirmed by **high-resolution analysis** of this ion. On the scale $C = 12.0000$, other elements have a non-integer relative atomic mass (*e.g.* $H = 1.0078$ and $O = 15.9949$). Consequently, the mass of an ion measured to four or more decimal places corresponds to a unique elemental composition. The isotope distribution of the halogens chlorine and bromine is also particularly helpful in the detection of these elements. Chlorine possesses two isotopes, ^{35}Cl and ^{37}Cl, in an approximate ratio of 3:1, and those of bromine, ^{79}Br and ^{81}Br, occur in a ratio of approximately 1:1. This leads to multiple ions of an intensity ratio reflecting the presence of these isotopes.

The extent to which **fragmentation** information can be used in the preliminary stages of the characterization of a compound depends to a considerable extent on the nature of the problem. If the compound being examined is one of a closely related series, the fragmentation pattern may possess readily identifiable ions. An example of this was in the structure elucidation of dehydrogeosmin (**2.3**), which was obtained from the flowers of a cactus, *Rebutia marsoneri*. The base peak (**2.4**) in the mass spectrum arose by a retro-Diels–Alder fragmentation. Although only 6 µg was available for analysis, this mass spectrum was sufficient to suggest a target structure for a confirmatory synthesis. Certain classes of compound fragment in predictable ways, polypeptides for example breaking at the peptide bond. Even if this is not the case, it is usually possible to find some helpful fragments based on the loss of a methyl group ($M - 15$), water ($M - 18$), carbon monoxide ($M - 28$) and their various combinations. The exact mass of the base peak may also be particularly informative.

M^+ 180   *m/e* 126
2.3 Dehydrogeosmin   **2.4**

2.2.2 Infrared Spectroscopy

The absorption of energy by a molecule in the IR region of the spectrum brings about vibrational and rotational changes.

The IR spectrum may afford useful information on the presence of particular **functional groups**. Although, ideally, spectra are determined in solution in solvents such as carbon tetrachloride (tetrachloromethane) to minimize intermolecular interactions, in practice this is often limited by the solubility of the natural product and by the fact that no solvent is completely transparent throughout the useful regions of the spectrum. The use of Nujol (liquid paraffin) as a mull or of a potassium bromide disc is a compromise. However, many spectroscopic correlations were obtained for solution spectra. Because a Nujol mull retains intermolecular effects, for example hydrogen bonding, which perturb the absorption of functional groups in adjacent molecules, the finer details of IR correlations are not always applicable. In the case of potassium bromide discs, salt formation with alkaloids has sometimes been observed.

Absorption in the 3200–3650 cm^{-1} region of the spectrum is characteristic of a **hydroxyl group**. Sometimes two bands are seen arising from the presence of hydrogen bonded and "free" hydroxyl groups. Whereas the former is a broad absorption, the latter is often sharp (3550–3650 cm^{-1}). The inter- and intramolecular hydrogen bonding of carboxylic acids serves to produce characteristic very broad absorption in the region 2500–3200 cm^{-1}. **Amino** groups absorb in the same region of the spectrum (3300–3500 cm^{-1}) and amides may produce several bands in the 3050–3200 cm^{-1} region associated with hydrogen bonding effects. Amides and carboxylic acids also have significant carbonyl absorption.

The region of the spectrum around 3000 cm^{-1} is associated with C–H absorption. Acetylenic C–H, relatively rare in natural products, appears at 3300 cm^{-1}. Absorption between 3010 and 3090 cm^{-1}, associated with arenes and alkenes, and between 2700 and 2900 cm^{-1}, associated with aldehydic C–H, is sometimes detectable, but there are other better spectroscopic methods (*e.g.* NMR) for identifying these groups. The same applies to the weak absorption associated with acetylenes (alkynes) and nitriles around 2200 cm^{-1}.

The region between 1650 and 1850 cm^{-1} is characteristic of the **carbonyl** group. Within this region, absorption due to various types of carbonyl group can be distinguished (see Table 2.1). Many of these useful correlations were established through the study of secondary metabolites. Unsaturation serves to lower the frequency of absorption, but increasing ring strain increases it. The deductions drawn for this area of the spectrum have obvious implications in the interpretation of the ^{13}C NMR spectrum, and hence these correlations are also shown in Table 2.1. Absorption between 1580 and 1680 cm^{-1} may be associated with **alkenes** and **arenes**. Since the intensity of IR absorption is approximately proportional to the square of the change of the dipole moment during a vibration, these absorptions are usually much weaker than those of a carbonyl group. Whereas aromatic rings absorb at 1500 and between 1580 and 1620 cm^{-1}, alkenes absorb between 1620 and 1680 cm^{-1}, depending on their degree of substitution. The interpretation of this region of the spectrum again has important implications in the NMR spectrum and for conjugated double bonds in the UV spectrum.

Table 2.1   IR absorption[a] and ^{13}C NMR signals of carbonyl compounds

	$IR\,(cm^{-1})$	$^{13}C\,(\delta)$		$IR\,(cm^{-1})$	$^{13}C\,(\delta)$
Ketones			Esters		
acyclic unconj.	1715	205–218	unconj.	1740	
acyclic conj.	1670	196–199	conj.	1720	169–176
cyclic 6-ring unconj.	1720	209–213	vinyl	1770	
cyclic 6-ring conj.	1680	199–201	phenyl	1765	
cyclic 5-ring unconj.	1745	214–220	Lactones		
cyclic 5-ring conj.	1715		6-ring	1740	170–178
alkyl aryl	1690	190–195	5-ring	1770	
diaryl	1660				
Aldehydes[b]			Acids		
unconj.	1730	200–205	unconj.	1715	166–181
conj.	1695	190–195	conj.	1680	
Quinones	1670	180–187	aryl	1690	

[a] The infrared absorption forms the centre of a range ($\pm 10\ cm^{-1}$). Carboxyl groups show broad OH absorption in the IR
[b] Aldehyde C–H appears at δ_H 9–10 ppm

The fingerprint region of the spectrum is the region below 1500 cm^{-1} which contains absorption from skeletal vibrations that are characteristic of individual compounds.

There are a number of useful absorptions that may be identified within the **fingerprint region**, including C–O absorption, particularly of esters. Different types of alkene (see Table 2.2) may be distinguished by their absorption between 650 and 1000 cm^{-1} and there are correlations associated with the different substitution patterns of aromatic rings. However, much clearer spectroscopic evidence for this is available from the ^1H NMR spectrum.

Table 2.2   IR absorption and ^1H NMR signals for alkenes

	$IR\,(cm^{-1})$	$^1H\,NMR\,(\delta)$
H_a \ / H_b C=C / \ H_c	1000–980 920–900	H_a 5.75–5.95 H_b 4.85–4.95 H_c 4.95–5.00
\ C=CH$_2$ /	900–880	4.60–5.00
H \ / H C=C / \	730–660	5.30–5.50

Table 2.2 *Continued*

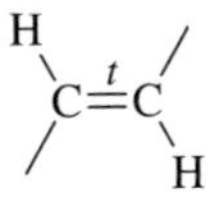 (structure: $C=C$ with H groups, *t*)	980–950	(note: different J values)
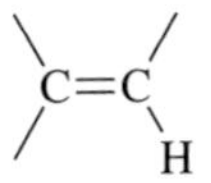 (structure: $C=C$ with H group)	850–790	5.50–5.70

2.2.3 Ultraviolet Spectroscopy

The UV spectrum provides a useful means of detecting **conjugated unsaturated chromophores** within a molecule such as polyenes, α,β-unsaturated ketones and aromatic compounds. Within particular families of compound the position of maximum absorption can reflect the degree of substitution of the chromophore. These correlations are summarized by the **Woodward–Fieser rules** for dienes and unsaturated ketones, and are given in Tables 2.3 and 2.4. Much of this structural information is also revealed by the ^{1}H and ^{13}C NMR spectra.

The absorption of energy by a molecule in the UV region of the spectrum brings about the promotion of electrons from one energy level to another. The easily observed absorptions involve π--π^* transitions.

The chromophore is the portion of the molecule responsible for absorbing the light.

Table 2.3 Woodward–Fieser rules for calculating UV maxima of dienes

Transoid dienes	nm	Cisoid dienes	nm
Parent: 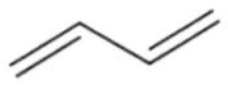 (structure)	214	 Parent: (structure)	253
Increment to be added for each			
Alkyl substituent	5		5
Extra conj. double bond	30		30
Exocyclic component	5		5
Calc. λ_{max}	——		——

UV absorption is characteristically broad, and the shape of the absorption curve, including the presence of shoulders on a band, may be diagnostic. This is found in aromatic absorption. It was also particularly helpful in detecting the presence of unstable polyacetylenes in culture broths of *Basidiomycete* fungi. The UV absorption of a common fungal metabolite, **ergosterol (2.5)**, is characteristic of its ring B diene. UV absorption is associated with the chromophore and not the whole molecule. Thus UV absorption would not distinguish between ergosterol

Table 2.4 Woodward–Fieser rules for calculating UV maxima of α,β-unsaturated ketones

	nm	^1H NMR signals (δ)
Parent: 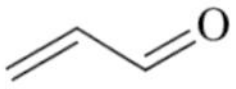	215	
	Increment to be added for each	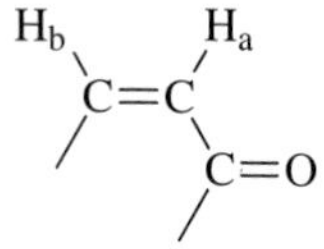
Extra conj. double bond	30	
Alkyl substituent α	10	
β	12	
γ, δ	18	$H_a = 5.80$
OH or OR	35	$H_b = 6.0\text{–}6.2$
Exocyclic component	5	
Extra C=O	0	
Calc. λ_{max}	——	

and ergosteryl esters, which often co-occur. In a useful application of UV, cleavage of **vitamin B$_1$** (thiamine, **2.6**) gave two fragments (Scheme 2.1), a pyrimidine (**2.7**) and a thiazole (**2.8**), which were recognized by their spectra and then synthesized.

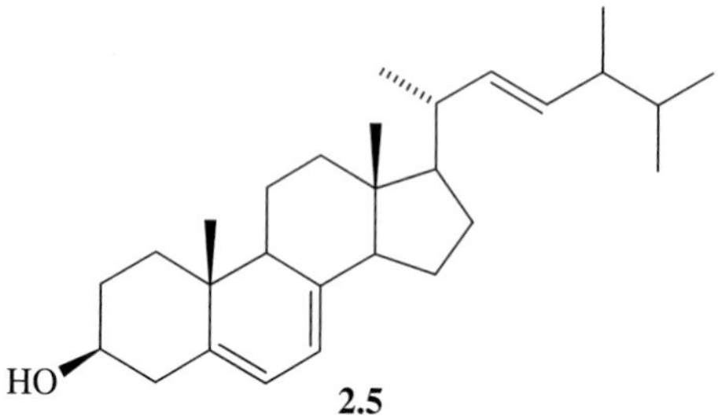

2.5

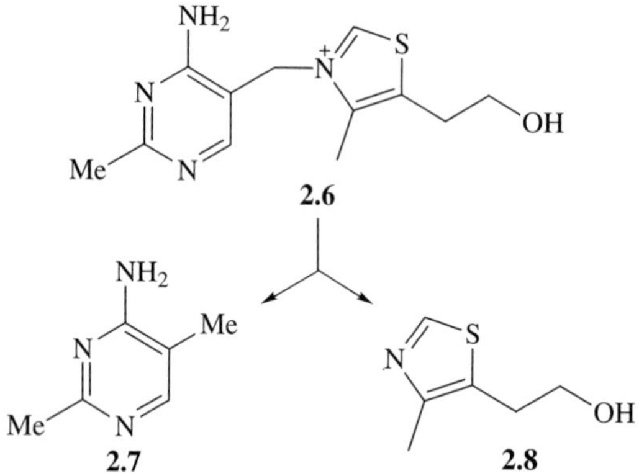

2.6

2.7

2.8

Scheme 2.1 Cleavage of vitamin B$_1$

The presence of **auxochromes**, particularly phenolic and carbonyl groups, modify aromatic chromophores. This can be particularly helpful in the identification of chromones and flavones. The shifts in the absorption of **phenols**, arising from the addition of a drop of alkali to produce the phenolate anion, are quite helpful. Various complexing agents also produce useful diagnostic changes in the UV spectrum. The UV spectrum may be a summation of chromophores from different parts of a polyfunctional molecule, and this should be considered in the light of deductions drawn from other spectroscopic methods and chemical degradation. An example where this was important was in the structural studies on the antibiotic **griseofulvin** (see Chapter 4). This antibiotic contains both a β-methoxy-α,β-unsaturated ketone and an acetophenone. Destruction of the former left the aromatic chromophore intact, and it was then identified.

An auxochrome is a substituent on the chromophore which modifies the absorption of a chromophore.

2.2.4 NMR Spectroscopy

The NMR spectra can be the most rewarding of the spectroscopic techniques in terms of structure elucidation. At the stage of the preliminary characterization the problem is often how to handle the wealth of data available and to distinguish the significant structural features. In tabulating data from the proton NMR spectrum it is worth remembering that the **chemical shift** provides information on the **chemical environment** of the proton, the **multiplicity** of a signal on the **relationship** with neighbouring protons and the **integral** gives the **relative number** of protons contributing to a signal. The columns of any table should reflect these aspects of the spectrum. The major regions of the proton NMR spectrum are given in Table 2.5.

Table 2.5 Approximate ranges of ^{1}H NMR signals

	δ		δ
Aliphatic C–H	0.5–2.0	Alkene C–H	4.5–6.5
C–H adjacent to C=X	1.75–2.5	Arene C–H	6.5–8.5
C–H adjacent to N	2.1–2.9	Aldehyde C–H	9.0–10.0
C–H adjacent to O	3.0–4.5	Carboxyl O–H	10.0–13.0

From the tabulation of data, it is helpful to determine the number and type of methyl group, *e.g.* C–CH$_3$, X=C–CH$_3$, N–CH$_3$ and O–CH$_3$; the number of protons on carbon atoms bearing oxygen functions, *e.g.* CH$_n$–O, O–CH$_n$–O and CH=O; and the number and type of alkene and aromatic protons. This will then leave a number of proton resonances, particularly within the region δ 1.0–2.5 (the methylene envelope), which may not be assigned on the first inspection. At this stage, shaking the NMR solution with deuterium oxide may identify the signals for readily exchangeable (OH and NH) protons. This may also serve to clarify or simplify some of the C<u>H</u>(OH) signals. It is possible to identify adjacent systems from common coupling constants. Some characteristic spin systems (*e.g.* the quartet and triplet of an ethyl group) may be apparent. Examination of the multiplicity of arene and alkene proton resonances may be used to determine their substitution patterns. Simple decoupling experiments to confirm these relationships may be carried out at this stage of the structure elucidation. It is important to do a "proton" count to check that this agrees with the molecular formula. This may reveal acidic protons whose low-field resonances have not been found in the initial spectrum.

The systematic analysis of the carbon-13 NMR spectrum provides useful structural information. Two sets of spectra are usually obtained. The first is the **proton noise decoupled** spectrum in which all the C–H couplings are removed. This spectrum contains the **chemical shift** information. The second set of spectra give the **number of hydrogens** attached to each carbon. Using the **DEPT** (distortionless enhanced polarization transfer) pulse sequence, the carbon signals arising from a methylene appear with one sign, usually negative, while methyls and methines appear with the opposite sign. Quaternary carbon atoms do not appear, although other sequences can be used to enhance them. On older instruments the off-resonance proton-decoupled spectrum fulfilled the same purpose. In this spectrum, a residue of the carbon–proton coupling was retained, and methyl signals appeared as quartets (q), methylenes as triplets (t), methines as doublets (d) and quaternary carbons as singlets (s). The results of attached hydrogen experiments are sometimes quoted as q, t, d and s, even though they may have been established using the DEPT sequence. The data are then tabulated and a number of assignments may be made. Some characteristic chemical shift ranges are given in Table 2.6. In particular, aliphatic carbon atoms bearing oxygen functions and the carbon atoms of alkenes, arenes and carbonyl groups may be readily assigned. Within these groups, some more detailed information may be obtained by the use of empirical **additivity** rules. These relate the substituents attached to a carbon and its neighbours to the chemical shift. Although these work relatively well for simple molecules bearing one functional group, polyfunctional molecules bearing several groups which may interact, *e.g. ortho* substituents on an aromatic ring, give poorer agreement. Some useful correlations exist between the nature of a carbonyl carbon and its chemical shift. These are given in Table 2.1, along with the IR data.

Table 2.6   Approximate ranges of ^{13}C NMR signals

	δ		δ
sp^3 C–C	10–50	–C≡C–	75–100
sp^3 C–N	25–75	>C=C<	105–160
sp^3 C–O	50–90	Arene C	110–155
sp^3 C–hal	35–65	C=O	see Table 2.1

In examining the carbon-13 data, it is important to do a carbon count. Some signals, particularly carbonyl signals with long relaxation times, do not always appear, and recording parameters may have to be changed to detect them. This may also reveal aspects of local symmetry within a molecule, such as a *para*-disubstituted aromatic ring in which some carbon atoms have an identical environment and hence identical chemical shifts and appear as one signal. It is also possible, using the attached hydrogen test, to check the number of hydrogen atoms that have been located.

When tabulating the data, various proton–carbon connectivities may be apparent. These may be confirmed in later stages of the structure elucidation by using **two-dimensional** spectra, *e.g.* a 2D-COSY (COrrelated SpectroscopY) spectrum in which the proton signals are plotted against the carbon signals.

In two-dimensional spectra there are two frequency axes and two spectra, *e.g.* the ^1H and ^{13}C spectra of a compound are plotted against each other. The cross peaks reveal correlations between signals in the two spectra.

Having completed the spectroscopic analysis, the number of degrees of unsaturation which have to be accomodated in the structure will be clear. Implicit in the molecular formula is the total number of double bond equivalents (see p. 36). The difference between these two gives the number of rings, each ring being equivalent to two hydrogen atoms.

2.3   Simple Chemical Derivatives

If there is sufficient material available, a number of simple derivatives may be prepared at this stage of the structure determination. The changes in the spectra which accompany these reactions may then provide further structural information. For example, if **hydroxyl** groups have been detected, acetylation with acetic (ethanoic) anhydride in pyridine can be informative. Apart from changes in the chemical shift of the CH–O–R resonances, the number of acetyl (CH$_3$C=O) groups that are introduced may indicate the number of primary or secondary hydroxyl groups that are present. Mild acetylation conditions do not normally acetylate a tertiary hydroxyl group or some hindered secondary hydroxyl groups. Hence the IR spectrum should be checked for any remaining hydroxyl absorption. Oxidation of a hydroxyl group to a carbonyl group may, from the carbonyl absorption in the IR and from the position of the carbon-13 carbonyl signal, reveal the size of the ring bearing the hydroxyl group. Changes in the UV absorption may reveal the proximity of the hydroxyl group to unsaturation. If a **carboxylic acid** has been detected, methylation with diazomethane should give a methyl ester. The appearance of new methoxyl signals should correspond with the number of carboxyl groups. Esters are often more soluble in organic solvents and are easier to purify than the parent acids.

The presence of an **alkene** may be confirmed by hydrogenation. Quantitative microhydrogenation can be carried out on quite a small scale (25 mg or less). Since one mole of hydrogen occupies 22.4 dm^3 at STP, it is relatively easy to measure the quantities of hydrogen that are absorbed by as little as 0.1 mmole (2.24 cm^3 per double bond). The changes in the spectra may reveal those functional groups that are close to the double bond. Before spectroscopic methods were available, microhydrogenation, peracid titration and iodination were methods that were used quantitatively to establish the number of double bonds in a molecule. Microscale ozonolysis, and determination of the formaldehyde (methanal) that was formed through its crystalline dimedone derivative, was a useful method for establishing the presence of a $=CH_2$ group.

The hydrolysis of **esters** and the cleavage of glycosides can also be useful preliminary chemical steps. However, there are a number of instances in which the methods that were used were too vigorous and led to rearrangements that confused the structure elucidation.

At this stage it is usually clear to what class of natural product the compound belongs and a working structure may have emerged. Study of the literature may reveal a possible relationship to a known compound, which can then be established experimentally. If this is not the case, the next phase of structure elucidation aims at establishing the underlying carbon skeleton.

2.4 The Determination of the Carbon Skeleton

In determining the carbon skeleton, the underlying strategy is one of structural simplification to reveal identifiable fragments which may then be linked together. Although the roles of spectroscopic and chemical methods in the elucidation of the carbon skeleton are interwoven, it is convenient to discuss them separately because the chemical methods evolved first. The chemical methods involved **degradations** to reveal **identifiable fragments**, the structures of which might be confirmed by synthesis. Spectroscopic studies seek to identify fragments which might then be linked together, *i.e.* to establish **connectivities**.

2.4.1 Oxidative Degradation

The vigorous oxidation of natural products by reagents such as chromium(VI) oxide, potassium manganate(VII) and concentrated nitric acid, followed by the identification of the often acidic products, played a major role in classical degradative chemistry. However, much of the information is now available from less destructive means. For example, in the past, oxidation with chromium(VI) oxide, steam distillation and identification of the number of moles of acetic acid which were formed, was the basis of the Kuhn–Roth determination of *C*-methyl groups. This information is now available from the ^{1}H NMR spectrum.

Aromatic rings often survived these oxidations intact, and hence the isolation of various aromatic polycarboxylic acids provided evidence for the substitution pattern of aromatic rings. Vigorous oxidation of the alkaloid **papaverine** (**2.9** in Scheme 2.2) gave firstly papaveraldine (**2.10**) and then dimethoxyisoquinoline-1-carboxylic acid (**2.12**), pyridine-2,3,4-tricarboxylic acid (**2.13**), metahemipinic acid (4,5-dimethoxybenzene-1,2-dicarboxylic acid, **2.15**) and veratric acid (3,4-dimethoxybenzoic acid, **2.16**). The identification of these fragments, together with the alkaline fission of the C_{20} alkaloid to a C_{11} base, 6,7-dimethoxyisoquinoline (**2.11**), and a C_9 ether, 4-methylveratrole (3,4-dimethoxymethylbenzene, **2.14**), was used in locating the position of the methoxyl groups in the alkaloid. Today, the presence and substitution pattern of aromatic rings would be revealed by the ^{1}H and ^{13}C NMR spectra.

Papaverine is one of the major constituents of the opium poppy.

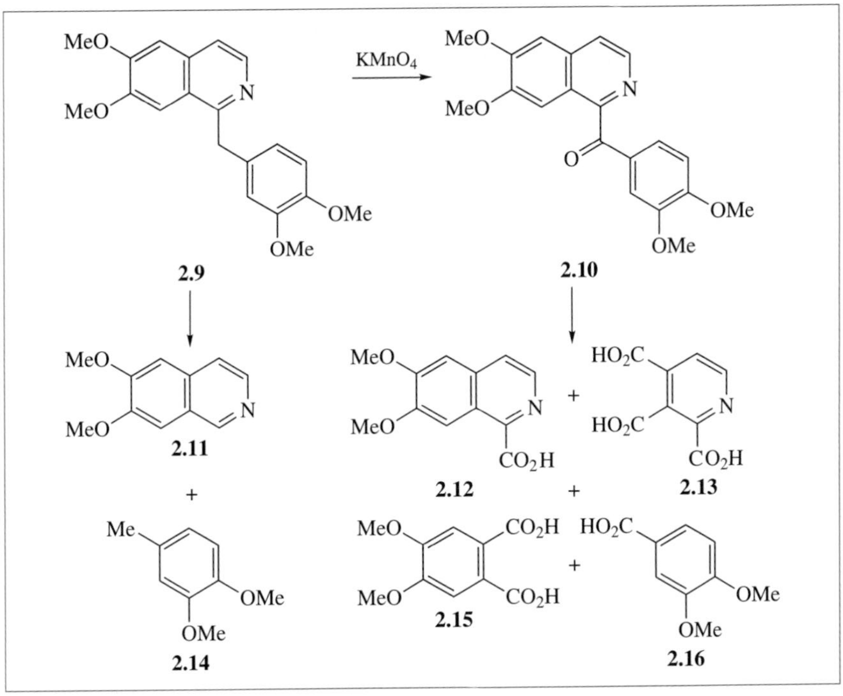

Scheme 2.2 Degradation of papaverine

α-Pinene is a major constituent of oil of turpentine. The elucidation of its structure played a pivotal role in the study of the monoterpenoids.

The stepwise oxidation of natural products with careful analysis and characterization at each step provided a useful strategy. In this context, an alkene often provided the initial site of attack, since it may be hydroxylated and cleaved or the allylic position may be attacked. The oxidative degradation of the monoterpene **α-pinene** (**2.1**; see Scheme 2.3) illustrates this. Mild oxidation with potassium manganate(VII) gave a glycol and, under more vigorous conditions, a C_{10} acid, pinonic acid (**2.17**). The latter, typically for a methyl ketone, underwent a **haloform** reaction with sodium bromate(I) to give bromoform (tribromomethane) and a C_9 dicarboxylic acid, pinic acid (**2.18**). Since there was no loss of carbon in the first oxidation, the original double bond was trisubstituted and bore a methyl group. This acid was brominated and converted with barium hydroxide to an α-hydroxy acid, which was in turn cleaved with chromic(VI) acid to a C_8 dicarboxylic acid, *cis*-norpinic acid (**2.21**). Hence one of the carboxyl groups of pinic acid was attached to a methylene. On heating this dicarboxylic acid with acetic anhydride it readily formed an internal anhydride, indicating a *cis* relationship between the two carboxyl groups. The dicarboxylic acid was isomerized with dilute hydrochloric acid to a *trans* isomer, *trans*-norpinic acid. This dimethylcyclobutanedi-carboxylic acid was then synthesized to confirm its structure. A relationship with the monocyclic monoterpene α-terpineol (**2.19**) was established by treatment of pinonic acid (**2.17**) with mineral acid. This led to opening of the cyclobutane ring and the formation of homoterpenyl methyl ketone (**2.20**), a compound which had already been obtained from α-terpineol. This type of degradation was very important in establishing the relationship between the monoterpenes during the latter part of the 19th century and the early part of the 20th century.

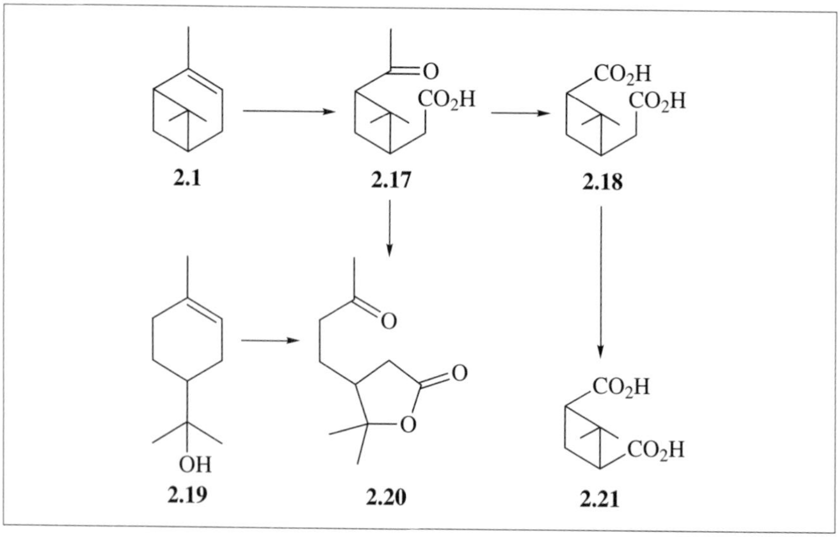

α-Terpineol is found in many essential oils, particularly pine oils. Its acetate is used in flavouring.

Scheme 2.3 Degradation of α-pinene

It is instructive to consider the changes in the spectra and the information which might have been obtained from them, had these methods been available at the time. In the oxidation of α-pinene (**2.1**) to pinonic acid (**2.17**), the alkene signals [δ_C 144.8 (C) and 116.4 (CH) and δ_H 5.17] disappeared. The IR spectrum of the oxidation product, [M^+ 184.110, $C_{10}H_{16}O_3$] showed two carbonyl absorptions (1731 and 1682 cm^{-1}) and broad hydroxyl absorption (3233 cm^{-1}), together with ^{13}C NMR signals at δ_C 178.6 and 207.4, characteristic of a carboxyl and a further carbonyl group. The ^{13}C NMR spectrum contained three methyl signals, two methylenes, two methines and a further quaternary carbon. The ^{1}H NMR spectrum contained two three-proton singlets at δ_H 0.83 and 1.28, assigned to two isolated methyl groups, together with a three-proton singlet at δ_H 2.03, typical of a methyl ketone. There was a one-proton double doublet at δ_H 2.84 (J = 7.9 and 9.8 Hz) which was coupled to a two-proton multiplet at δ_H 1.90, suggesting the presence of the group $CH_3C(O)CHCH_2$. There was a three-proton multiplet at δ_H 2.28 which was in the range expected for protons adjacent to a carboxyl (CH_2CO_2H). Thus even a cursory inspection of the spectra from this single oxidation reveals considerable structural information.

The presence of a ketone in a ring system has provided the starting point for some structurally informative ring cleavage reactions. Oxidation of the alkene of the monoterpene β-pinene (**2.2**), either with potassium manganate(VII) or by ozonolysis, gave the C_9 ketone nopinone (**2.22**, Scheme 2.4). Hence the double bond was an exocyclic methylene (=CH_2). Today this information would be obtained from the ^{1}H, ^{13}C NMR and IR spectra [δ_H 4.6 (2H); δ_C 106.0 (CH_2), 151.8 (C); ν_{max} 1641, 873 cm^{-1} (=CH_2)]. Further oxidation of the ketone with nitric acid led to ring opening and the formation of homoterpenylic acid (**2.23**), linking this compound with the α-terpineol series.

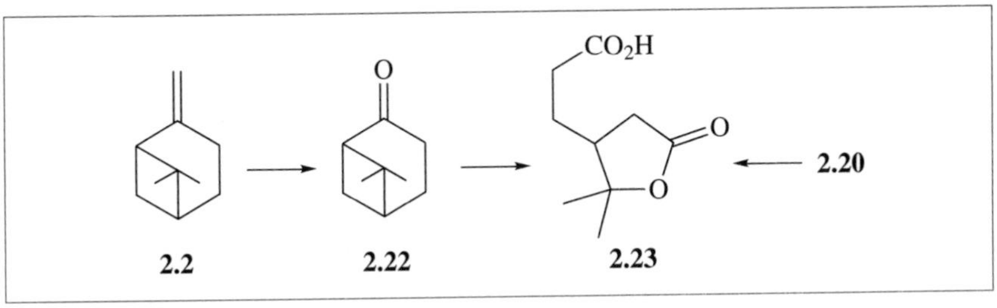

Scheme 2.4 Degradation of β-pinene

The oxidative cleavage of rings to dicarboxylic acids led to a chemical method for establishing the size of rings. In 1907, Blanc formulated a rule which stated that, if two carboxyl groups occupy the 1,5-positions as in a dicarboxylic acid arising from a cyclopentanone, the dicarboxylic acid on being heated with acetic anhydride is converted into an anhydride. In contrast 1,6- and higher dicarboxylic acids, arising from cyclohexanones and larger rings, give cyclic ketones with the loss of carbon dioxide. This oxidative degradation provided evidence for the size of ring A of the steroids, although it gave what turned out to be misleading evidence for the size of the more heavily substituted rings B and C. The size of a ring containing a carbonyl group is now established by IR and ^{13}C NMR spectroscopy (see Table 2.1).

2.4.2 Reductive Degradation

One enantiomer of myrtenol occurs in sweet orange oil and the other is found in *Cyperus articulatus*. In some parts of the world the dried tubers of this plant are used to perfume clothes.

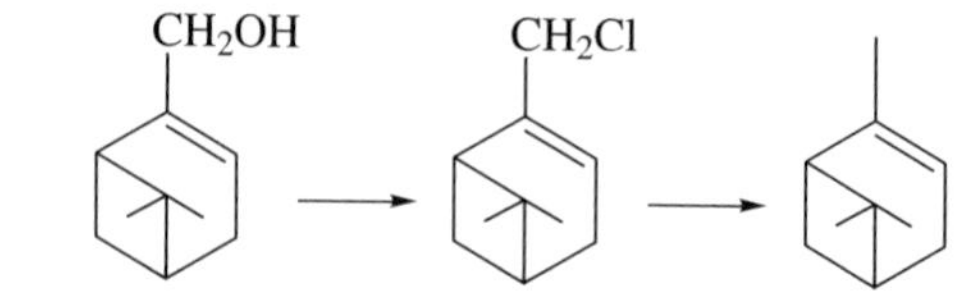

Many secondary metabolites belong to one of several common series and differ from each other merely in the position of hydroxyl groups on this carbon skeleton. One strategy which recognizes this is to remove the oxygen functions in a stepwise manner to obtain a hydrocarbon which may then be identified. For example, **myrtenol** (**2.24**) and the corresponding aldehyde, myrtenal, occur in a number of essential oils. Treatment of myrtenol with phosphorus pentachloride gave the chloro compound **2.25**, which was in turn converted to α-pinene (**2.1**) by reduction with sodium in alcohol (Scheme 2.5).

Scheme 2.5   Interrelationship of myrtenol and α-pinene

CH$_2$OH   CH$_2$Cl

2.24 Myrtenol   2.25   2.1

2.4.3 Dehydrogenation to Aromatic Compounds

Dehydrogenation was developed as a structural method in the 1920s by Ruzicka and had a considerable impact on the elucidation of the structures of polycyclic sesqui-, di- and triterpenes.

Dehydrogenation, particularly with **sulfur** or **selenium** to form an aromatic compound, has provided a powerful means of **structural simplification**. The effect of converting sp^3 centres to sp^2 centres is to remove problems of stereochemistry from consideration at this stage in the structure elucidation. The identity of the aromatic compounds was then established by synthesis. The synthesis of aromatic compounds involves the introduction of substituents based on well-established rules of aromatic substitution. The UV spectra of the different classes of aromatic compound are distinctive and can aid their identification.

An example of this is provided by the dehydrogenation of **abietic acid** (**2.26**), a major constituent of pine resin, which gave the phenanthrene retene (**2.27**, Scheme 2.6). This was then synthesized. In this hydrocarbon the carbon skeleton of abietic acid is retained with the loss of only two peripheral carbon atoms. The sesquiterpenoids were at one time grouped in terms of the naphthalene and azulene hydrocarbons obtained from them on dehydrogenation.

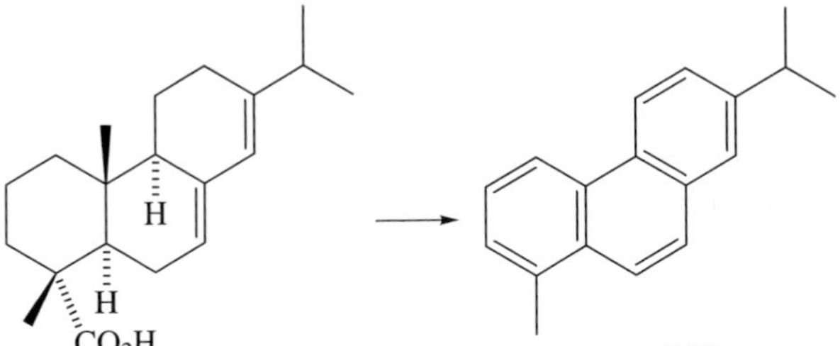

Scheme 2.6 Degradation of abietic acid

In these examples, quaternary methyl groups were expelled, but under milder dehydrogenation conditions such groups may be retained.

On dehydrogenation, compounds containing carbonyl groups give rise to phenols, which were often purified as their methyl ethers. This provided a useful means of locating the oxygen function on the carbon skeleton. An example came from the determination of the structure of allogibberic acid (**2.29**), which is a degradation product of the plant

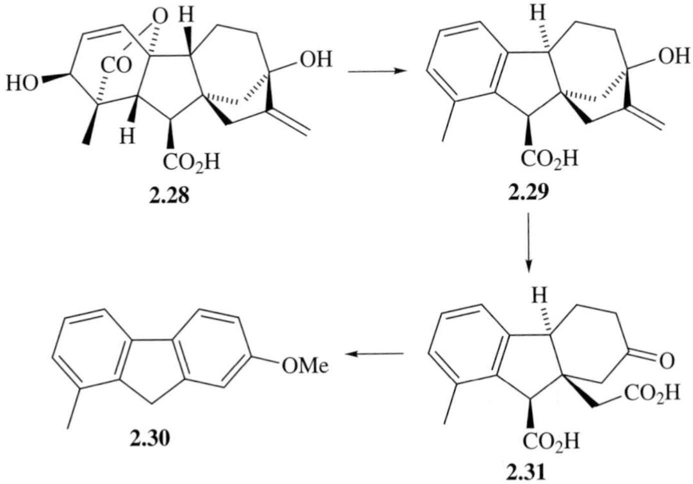

Scheme 2.7 Degradation of gibberellic acid

The structure of gibberellic acid was studied during the 1950/60s at a time of transition between the different roles of chemical and spectroscopic methods in structure determination.

hormone **gibberellic acid** (**2.28**, Scheme 2.7). Ring D was opened by ozonolysis to give a keto acid (**2.31**) in which the carbonyl group marked the site of the tertiary hydroxyl group. Dehydrogenation of the keto acid gave a methoxyfluorene (**2.30**), which was synthesized. Note the progressive structural simplification in this degradation and in particular the elimination of stereochemical problems.

Another strategy for locating the site of a carbonyl group uses a Grignard reaction with methylmagnesium iodide to introduce a methyl group at the site of the carbonyl group, so "marking" that position. Dehydrogenation then gives an aromatic compound which contains an additional methyl group when compared to that derived from the parent secondary metabolite. Such a strategy was used to locate the position of a carbonyl group in the diterpenoid **ketomanoyl oxide** (**2.32**, R = O). The carbon skeleton of manoyl oxide (**2.32**, R = H$_2$) was established by the isolation of 1,2,5-trimethylnaphthalene (**2.33**, R = H) and 1,2,8-trimethylphenanthrene on dehydrogenation (Scheme 2.8). When ketomanoyl oxide (**2.32**, R = O) was treated with methylmagnesium iodide and the product dehydrogenated, 1,2,5,7-tetramethylnaphthalene (**2.33**, R = Me) and 1,2,6,8-tetramethylphenanthrene were obtained. The extra methyl group marked the position of the carbonyl group in the original ketomanoyl oxide.

Manoyl oxide and ketomanoyl oxide are obtained from the wood oil of the New Zealand silver pine, *Dacrydium colensoi*. A more highly oxidized relative, forskolin, has been obtained from an Indian medicinal plant, *Coleus forskolii*, and has attracted interest because of its action in lowering blood pressure.

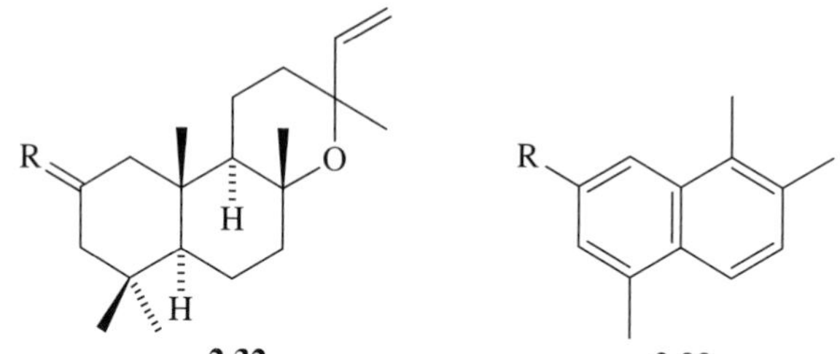

Scheme 2.8 Dehydrogenation of ketomanoyl oxide

2.32   2.33

Worked Problem 2.2

Q Eremophilone is a sesquiterpenoid ketone which was obtained from the Australian tree *Eremophila mitchelli*. On reduction it gives an alcohol which on dehydrogenation with selenium gave 7-isopropyl-1-methylnaphthalene. When the ketone was reacted with methylmagnesium iodide and the resultant tertiary alcohol dehydrogenated, the product was 7-isopropyl-1,5-dimethylnaphthalene. What is the location of the carbonyl group on the carbon skeleton?

A   The aromatic hydrocarbons are:

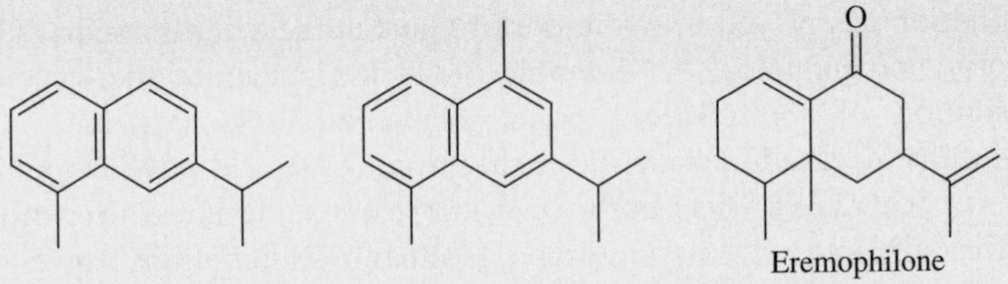

Eremophilone

These naphthalenes have been drawn to show their relationship to the eremophilane carbon skeleton. The position of the carbonyl group is revealed by the position of the additional methyl group. Note that the dehydrogenation led to the loss of one carbon atom. The final structure of eremophilone was interesting because it does not follow the isoprene rule and a rearrangement has taken place during its biosynthesis.

2.4.4 The Removal of Nitrogen

A nitrogen atom is a site of reactivity in an organic molecule, and can be the starting point for informative degradations. The best known of these is the **Hofmann elimination** of quaternary ammonium salts (Scheme 2.9). The amine is first methylated with iodomethane to form the quaternary ammonium salt. This salt is then heated with moist silver oxide to bring about the elimination and form an alkene. The number of N-methyl groups that are introduced distinguishes between a primary, secondary or tertiary amine in the starting material. Not only does this reaction have the

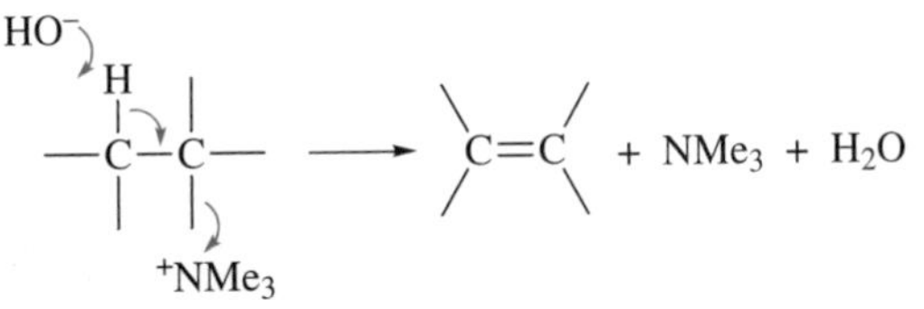

Scheme 2.9   The Hofmann elimination

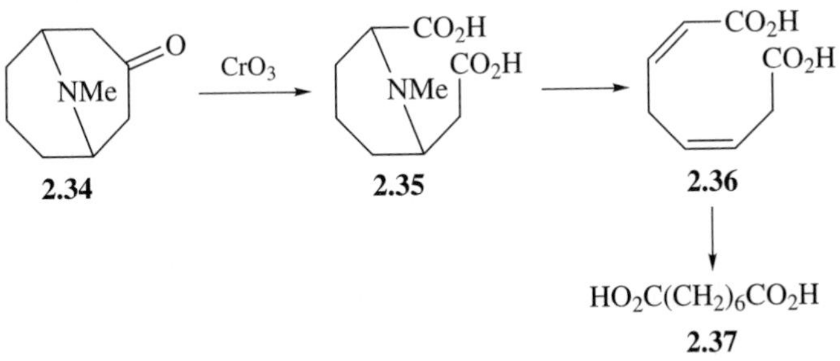

Scheme 2.10   Degradation of pseudopelletierine

advantage of breaking a carbon–nitrogen bond, but it also generates an alkene which may form the basis for further degradations. The Hofmann elimination has played a very important part in the determination of the structure of alkaloids. An example of its use comes in the structure elucidation of the alkaloid **pseudopelletierine** (**2.34**, Scheme 2.10). Oxidation of pseudopelletierine with chromic(VI) acid gave *N*-methyl-granatic acid (**2.35**). This dicarboxylic acid retained the nine carbon atoms of the alkaloid. Two successive Hofmann eliminations gave the unsaturated acid (**2.36**), which was reduced to the known straight-chain C_8 dicarboxylic acid suberic acid (octandioic acid, **2.37**). The first synthesis of the interesting hydrocarbon cyclooctatetraene (**2.38**) started from pseudopelletierine (**2.34**) and used a sequence of Hofmann eliminations.

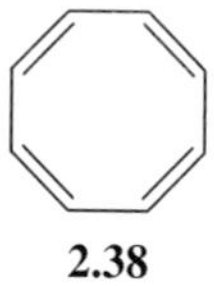

2.38

An alternative degradation, the **von Braun** reaction, converts the nitrogen into a good leaving group by reaction with cyanogen bromide. The cyanoammonium salt then undergoes a nucleophilic substitution by bromine. The course of these two degradations can be compared in the work on the alkaloid **hydrocotarnine** (**2.39**, Scheme 2.11). Another degradation, the Emde degradation, involves the reductive cleavage of a quaternary ammonium halide with sodium amalgam.

Hydrocotarnine is an alkaloid which has been obtained from the opium poppy and by degradation of another constituent, narcotine.

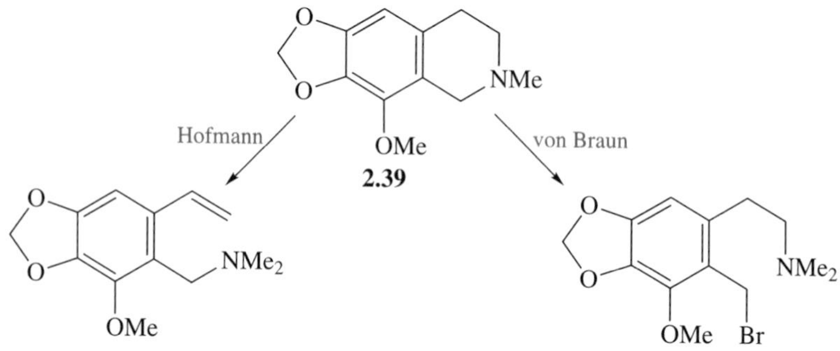

Scheme 2.11   Degradation of hydrocotarnine

Worked Problem 2.3

Q   The alkaloid derivative isothebaine methyl ether (**1**) gave, after two Hofmann degradations, the vinylphenanthrene **2**. Give a mechanism for these eliminations.

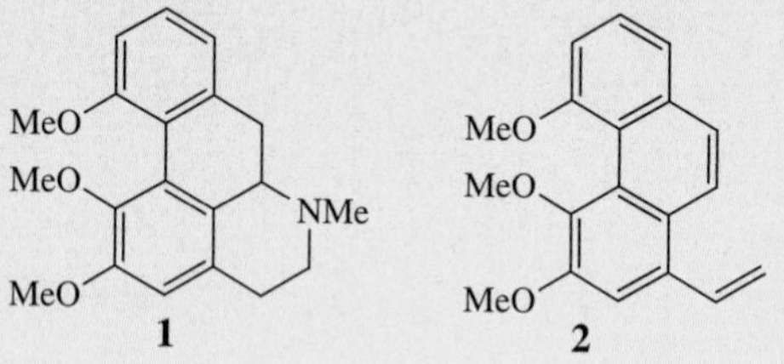

A A quaternary ammonium salt (**3**) is formed first and then, under the influence of a base, this undergoes an elimination reaction. In each case the β-hydrogen atom that is eliminated is adjacent to an aromatic ring and the extra double bond that is introduced is in conjugation with the ring:

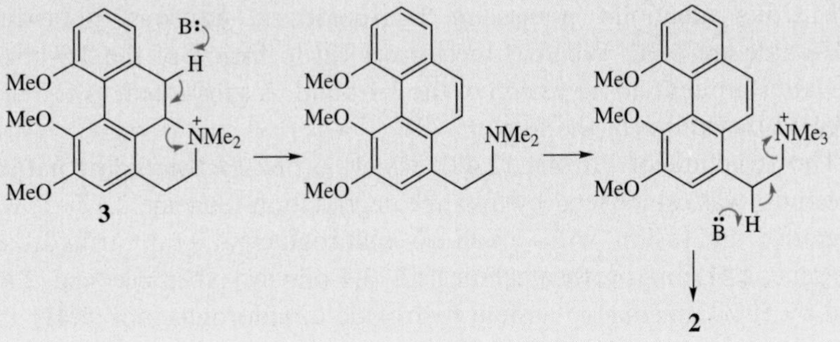

2.4.5 Alkaline Degradation

Many of the base-catalysed condensation reactions of synthetic organic chemistry are reversible. Retro-Claisen and retro-aldol reactions have played an important part in the degradation of natural products. The polyhydroxylated flavones include some of the major pigments of fruits,

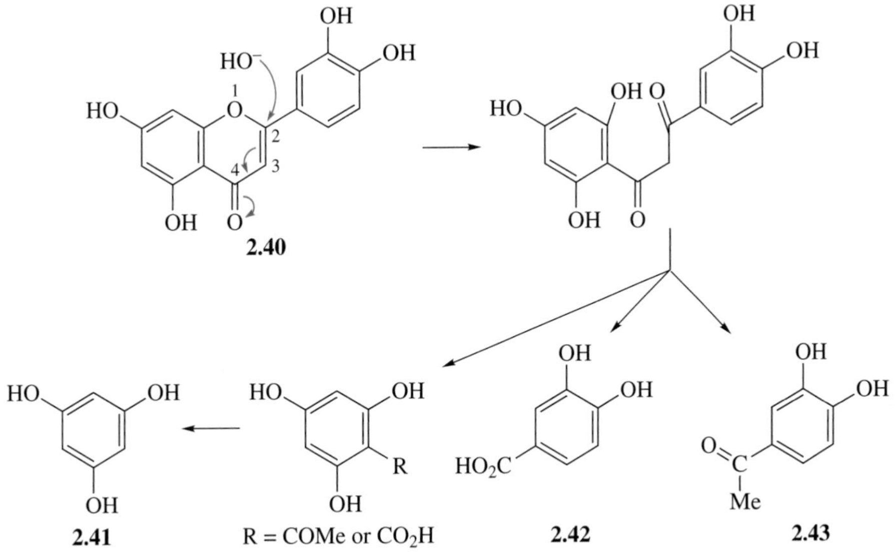

Scheme 2.12 Alkaline degradation of luteolin

flowers and the wood of some trees. Their structures were established chiefly by means of alkaline degradation followed by synthesis of the degradation products and then the flavanones themselves. Conjugate addition of a nucleophile to the chromone ring (*e.g.* **2.40**, Scheme 2.12) leads to a rapid hydrolysis to form a β-diketone. The final products obtained by alkaline fission of the diketone depend upon whether anhydrous ethanolic potassium hydroxide or aqueous potassium hydroxide are used. Whereas the former led to fission of the 2,3-bond, the latter brought about fission of the 3,4-bond. A substituent at C-3 also affected the course of the fission.

Luteolin is a yellow dyestuff which was used as a natural pigment and was obtained from the Mediterranean plant "dyer's weed" (*Reseda luteola*).

The structure of **luteolin** (**2.40**), which is one of the oldest natural dyestuffs, was established by alkaline degradation (Scheme 2.12). It was degraded by fusion with alkali to phloroglucinol (1,3,5-trihydroxybenzene, **2.41**) and protocatechuic acid (3,4-dihydroxybenzoic acid, **2.42**) and by 50% aqueous potassium hydroxide to phloroglucinol (**2.41**) and 3,4-dihydroxyacetophenone (**2.43**).

2.4.6 Acid-catalysed Degradations

Although acid-catalysed degradations have provided methods of structural simplification, they have also been the source of some confusion, particularly because of the propensity of natural products such as the terpenoids to undergo rearrangements under these conditions. Although not necessarily leading to the final structure, these rearrangements have nevertheless shed a fascinating light on the chemistry of the compounds concerned. The **Wagner–Meerwein** rearrangements were discovered in the context of structural studies on the bicyclic monoterpenes. The rearrangement of the sesquiterpenoid **santonin** (**2.44**, Scheme 2.13) to

The Wagner–Meerwein rearrangements were originally discovered in the conversion of α-pinene to bornyl chloride.

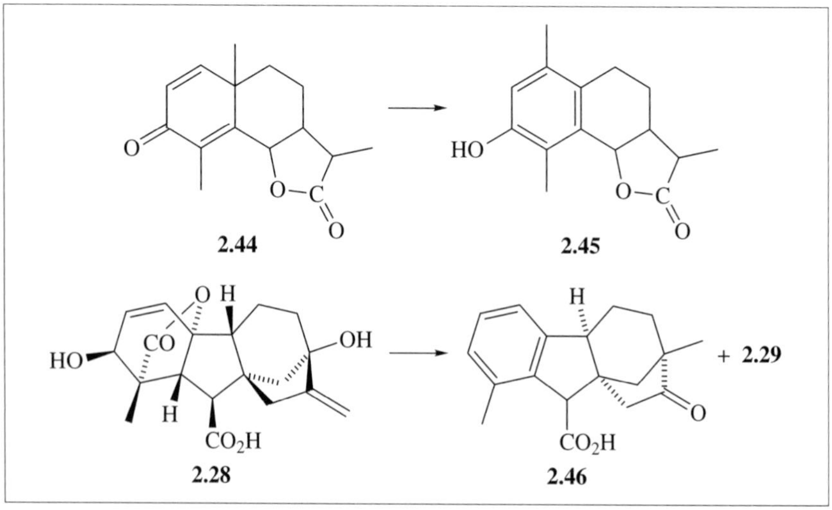

Scheme 2.13 Acid-catalysed degradations

desmotroposantonin (**2.45**) is an example (see Chapter 4). We have already noted the acid-catalysed cleavage of pinonic acid (**2.17**) to homoterpenyl methyl ketone (**2.20**), which played an important part in the inter-relationship of the monoterpenoids.

The elucidation of the structure of the plant hormone **gibberellic acid** (**2.28**) involved the study of its acid-catalysed decomposition products, allogibberic acid (**2.29**) and gibberic acid (**2.46**). These degradation products, while retaining the underlying carbon skeleton of gibberellic acid, had lost the complicated structural features associated with ring A. This was a very useful structural simplification.

2.5   Spectroscopic Methods in the Determination of the Carbon Skeleton

Over the last 30 years there has been a gradual change in emphasis in the role that spectroscopic methods have played in the determination of the carbon skeleton of a natural product. Spectroscopic methods were used to follow a chemical degradation; the chemical change came first and its spectroscopic consequences were then analysed, that is the strategy was essentially chemically based. However, particularly with the advent of two-dimensional NMR methods making connectivity studies easier, leading to the spectroscopic identification of fragments of a molecule, the strategy is increasingly based on spectroscopy.

2.5.1   NMR Spectroscopy in Determining the Carbon Skeleton

NMR spectroscopy dominates the determination of the structures of natural products. NMR has been used as an adjunct to chemical degradation since the late 1950s. Oxidative degradation, by introducing functionality into the molecule, produced a greater spread of ^1H and ^{13}C NMR signals, enabling groups adjacent to reactive centres to be identified. Furthermore, the use of spin decoupling techniques enabled adjacent protons to be recognized, and hence significant structural fragments could be identified.

Within a related series of compounds it is often possible to identify certain characteristic structural fragments from their spectroscopic fingerprints. For example, the gibberellin plant hormones (**2.28**) have a characteristic AB double doublet which is assigned to H-5–H-6.

^1H–^1H and ^1H–^{13}C correlations can be made using two-dimensional NMR studies. This has opened up new strategies of structure deter-mination which can be led by spectroscopy. In these studies, part or even the complete structure may be derived from internal evidence, with less reliance on analogies to other compounds. The homonuclear techniques are known by the acronyms of ^{1}H–^{1}H **COSY** (correlation spectroscopy), which reveals coupling information, and **NOESY**, which reveals nuclear Overhauser effect interactions (see p. 66). The heteronuclear COSY experiment reveals ^{1}H–^{13}C correlations, and may be used to identify both one-bond and, separately, long-range couplings. These correlations rely on the transfer of magnetization between coupled or sterically interacting nuclei.

With the variety of methods that are available, it is important to establish a strategy which does not waste material or instrument time. A typical strategy might be the following. Firstly, from the ^{13}C NMR spectrum and the attached hydrogen experiment (*e.g.* the DEPT spectrum, see p. 44), identify those carbons that are methyls, methylenes, methines and quaternary atoms. Secondly, from the ^{1}H–^{13}C heteronuclear chemical shift correlation spectrum, identify those proton signals that correspond to methylenes, *i.e.* those protons that might be geminally coupled. Thirdly, using the above information, identify those signals from the ^{1}H–^{1}H COSY spectrum that are vicinally coupled and relate these to their respective carbon atoms using the two-dimensional heteronuclear spectrum. The identification of vicinal couplings may need separate homonuclear spin decoupling or selective population transfer experiments to identify the magnitude of the coupling and distinguish between vicinal and long-range couplings. In this way it is possible to build up a series of related spin systems.

This strategy will yield some structural fragments which may reach a terminus in a quaternary carbon atom. These quaternary carbon atoms may then be identified by using a long-range ^{1}H–^{13}C correlation (**HMBC** = heteronuclear multiple bond correlation). This may provide the opportunity to link fragments that have been identified. A further way of linking structural fragments utilizes the nuclear Overhauser effect. This relies on a steric interaction rather than a bonding interaction and hence it can give different information on the relative position of protons in a molecule. The nuclear Overhauser effect is of considerable value in determining stereochemistry and we will therefore return to it later (see p. 66).

The keto diol **2.47** was obtained from the culinary fruit of the dill plant, *Anethum graveolens*. In establishing its structure a ^{1}H–^{1}H COSY NMR experiment linked H-1 to H-6 and H-7 and H-3 to H-4 and the ^{1}H–^{13}C heteronuclear correlation (HETCOR) linked these proton resonances to their respective carbon signals. The three-bond heteronuclear multiple bond correlation (HMBC) was then used to link H-1 and H-3 to the C-2 carbonyl group and H-4, H-9 and H-10 to the quaternary carbon at C-8. The links between these resonances led to the structure **2.47** for this compound.

geminal coupling   vicinal coupling

2.47

Summary of Key Points

1. A natural product may be characterized by physical, analytical, chromatographic and spectroscopic data.

2. This information will provide the molecular formula, the number of double bond equivalents, and lead to the identification of the functional groups (OH, C=O, C=C) and other salient structural features such as *C*-methyl, *N*-methyl or *O*-methyl groups.

3. The underlying strategy for establishing the carbon skeleton of a natural product involves structural simplification to reveal identifiable fragments.

4. Oxidative degradation, reduction to a parent hydrocarbon, dehydrogenation, elimination reactions and alkaline fission play an important role in establishing the carbon skeleton.

5. Spectroscopic methods, particularly NMR, play a major role in the identification of structural fragments. Two-dimensional correlation spectroscopy (2D-COSY) and heteronuclear multiple bond correlation spectroscopy (HMBC) NMR techniques have become important in relating groups of hydrogen and carbon atoms.

Problems

2.1. How might (i) a physical and (ii) a chemical method be used to distinguish between each of the following pairs:

(a)

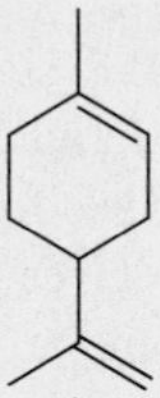

A   and   B

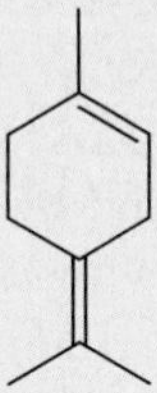

(b)

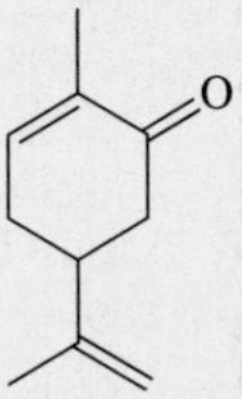

C   and   D

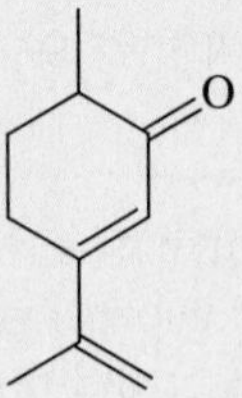

(c)

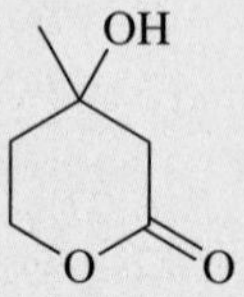

E   and   F

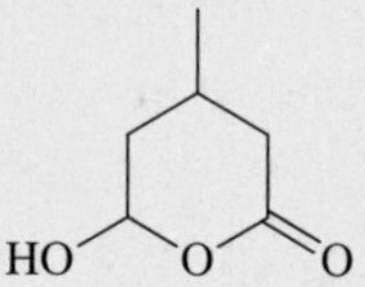

(d)

$$CH_2OH$$
$$|$$
$$CHOAc$$ and
$$|$$
$$CH_2OH$$
G

$$CH_2OAc$$
$$|$$
$$CHOH$$
$$|$$
$$CH_2OH$$
H

(e)

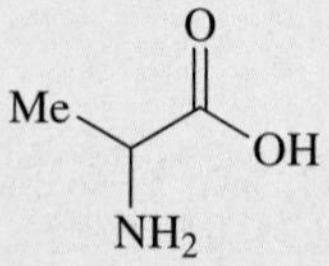

I   and   J

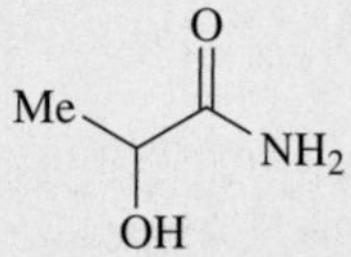

(f)

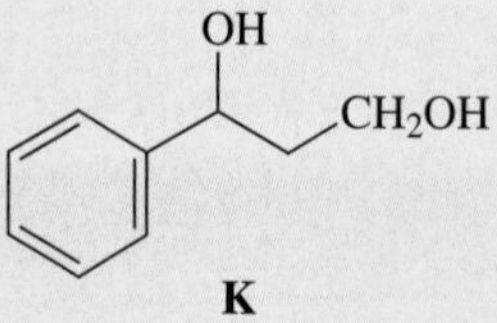

K   and   L

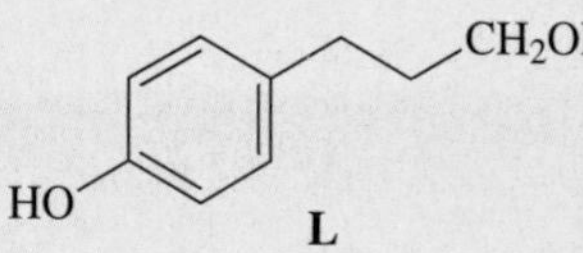

2.2. A volatile plant product **M**, $C_8H_{14}O$, has strong IR absorption at 1717 cm^{-1}. It possesses ^{1}H NMR signals at δ_H 1.65 (3H, s), 1.72 (3H, s), 2.15 (3H, s), 2.4 (4H, m) and 5.20 (1H, t). On ozonolysis, compound **M** gives, among other products, propanone, and on treatment with iodine and alkali it gives triiodomethane (iodoform). What structural information can be obtained from: (a) its molecular formula; (b) its IR spectrum; (c) its ^{1}H NMR spectrum; (d) its chemical reactions? (e) In the light of these answers, deduce a structure for compound **M**.

2.3. Hydrolysis of a natural product gave an aromatic compound **N**, $C_9H_{10}O_4$, which was soluble in aqueous sodium hydrogen carbonate. It had IR absorptions at 2700 (broad), 1690 and 1600 cm^{-1}. In the ^{1}H NMR spectrum there were signals at δ_H 3.85 (6H), 6.62 (1H) and 7.15 (2H) and a broad single-proton signal at δ_H 10.2 which disappeared when the sample was shaken with ^2H_2O. Careful inspection of the signal at δ_H 6.62 showed that it was a narrow triplet ($J = 1.5$ Hz), while that at δ_H 7.15 was a narrow doublet ($J = 1.5$ Hz). Suggest a structure for compound **N**, and account for the multiplicity of the signals.

2.4. A monocarboxylic acid **O**, $C_6H_6O_4$, has been obtained from a *Xestospongia* species. It possesses IR absorption at 3300 (broad), 1730 and 1700 cm^{-1}, and intense UV absorption at 217 nm. It possesses ^1H NMR signals at δ_H 2.78 (2H, d, $J = 7$ Hz), 5.40 (1H, m), 6.10 (1H, d, $J = 5$ Hz), 7.75 (1H, dd, $J = 2$ and 5 Hz) and 10.15 (1H, exchangeable with ^2H_2O). In a spin decoupling experiment, irradiation of the signal at δ_H 5.40 converted the signal at δ_H 2.78 to a singlet and the signal at δ_H 7.75 to a doublet, $J = 5$ Hz. When compound **O** was heated with sodium hydroxide and the solution carefully acidified, a dicarboxylic acid **P**, $C_6H_8O_5$, was obtained. Account for this observation and deduce a structure for **O**.

2.5. Sherry lactone, $C_6H_{10}O_3$, had IR absorption at 3500 and 1750 cm^{-1}. The ^1H NMR spectrum contained signals at δ_H 1.25 (3H, d, $J = 6.8$ Hz) and one-proton multiplets at δ_H 2.05, 2.26, 2.50, 2.70, 3.78 and 4.38. Analysis of the multiplet at δ_H 3.78 showed that it was a quartet ($J = 6.8$ Hz) of doublets ($J = 6.0$ Hz), while that at δ_H 4.38 was a triplet ($J = 7.0$ Hz) of doublets ($J = 6.0$ Hz). Irradiation of the signal at δ_H 3.78 converted the signal at δ_H 1.25 into a singlet and that at δ_H 4.38 into a triplet ($J = 7.0$ Hz). Oxidation of sherry lactone with pyridinium chlorochromate gave solerone, $C_6H_8O_3$, ν_{max} 1750, 1700 cm^{-1}. This oxidation product had ^1H NMR signals at δ_H 2.0 (3H, s), 2.1–2.7 (4H, m) and 4.65 (1H, t, $J = 7.0$ Hz). Propose structures for sherry lactone and solerone.

2.6. A sesquiterpenoid **Q**, with a structure that obeys the isoprene rule, has the formula $C_{15}H_{28}O$ and possesses IR absorption at 3500 cm^{-1}. On treatment with acetic anhydride, it gave a compound $C_{17}H_{30}O_2$ with IR absorption at 1730 cm^{-1}. Dehydrogenation of the sesquiterpenoid with selenium gave 7-isopropyl-1-methylnaphthalene. Treatment of the sesquiterpenoid with chromium(VI) oxide gave a compound, $C_{15}H_{26}O$, with IR absorption at 1700 cm^{-1}. When this compound was reacted with methylmagnesium iodide and the product was dehydrogenated with selenium,

7-isopropyl-1,2-dimethylnaphthalene was obtained. Suggest a structure for sesquiterpenoid **Q** and indicate possible dissections into the isoprene units in this structure.

2.7. A volatile component **R**, $C_8H_8O_2$, of the secretions of a beetle, *Phorocantha semipunctata*, had IR absorption at 3100, 1670, 1622 and 1580 cm^{-1} and UV absorption at 210 and 270 nm. The compound had ^{1}H NMR signals at δ_H 2.61 (3H, s), 6.75 (1H, d, $J = 8$ Hz), 6.95 (1H, d, $J = 8$ Hz), 7.42 (1H, t, $J = 8$ Hz), 10.25 (1H, s) and 11.98 (1H, s). The signal at δ_H 11.98 disappeared when the sample was shaken with ^2H_2O. Compound **R** formed a red 2,4-dinitrophenyl-hydrazone. It was obtained, along with a second product, compound **S**, when *m*-cresol (3-methylphenol) was treated with trichloromethane and sodium hydroxide. Account for these observations and deduce the formulae of the natural product **R** and the compound **S**.

2.8. Compound **T**, $C_6H_8O_2$, obtained from the insect *Sigara falleni*, had IR absorption at 2840, 2740, 1690, 1621 and 980 cm^{-1}. It gave a silver mirror on treatment with moist silver oxide, and it formed a bis(2,4-dinitrophenylhydrazone) derivative. The ^{1}H NMR spectrum showed signals at δ_H 1.17 (3H, t, $J = 7.3$ Hz), 2.74 (2H, q, $J = 7.3$ Hz), 6.78 (1H, dd, $J = 7.1$ and 16.6 Hz), 6.89 (1H, d, $J = 16.6$ Hz) and 9.85 (1H, d, $J = 7.1$ Hz). Identify the adjacent ^{1}H NMR signals from their coupling patterns and assign a structure to compound **T**.

2.9. A sesquiterpenoid **U**, $C_{15}H_{24}O$, had IR absorption at 1680 cm^{-1} and UV absorption at 239 nm. The ^{1}H NMR spectrum included signals at δ_H 0.90, 1.0 and 1.05 (each 3H, d), 1.80 (3H, s) and 5.2 (1H, s). On catalytic hydrogenation, compound **U** gave compound **V**, $C_{15}H_{26}O$, which had IR absorption at 1710 cm^{-1} and which lacked the UV absorption at 239 nm. Further reduction with sodium borohydride gave compound **W**, $C_{15}H_{28}O$, which had IR absorption at 3500 cm^{-1}. Dehydrogenation of **W** with selenium gave 4-isopropyl-1,6-dimethylnaphthalene. Reaction of **V** with methyl-magnesium iodide gave **X**, $C_{16}H_{30}O$, which on dehydrogenation with selenium gave 4-isopropyl-1,6,8-trimethylnaphthalene. What are the structures of compounds **U–X**?

3

The Location of the Functional Groups and the Molecular Stereochemistry

Aims

The object of this chapter is to outline the general strategy for establishing the position of the functional groups and the stereochemistry of a natural product. By the end of this chapter you should understand:

- The role of spectroscopic methods, particularly NMR spectroscopy, in establishing interrelationships between functional groups
- The use of chemical methods including specific oxidations, retro-aldol and retro-Claisen reactions and cyclization reactions which reveal interactions between functional groups and hence their relative stereochemistry
- The use of stereochemical correlations, asymmetric induction and spectroscopic methods in determining absolute stereochemistry

3.1 Introduction

The first two phases in the elucidation of the structure of a natural product lead to the identification of the functional groups and the structure of the carbon skeleton. Information often accumulates during these stages on the position and stereochemistry of functional groups. Consequently, it is often helpful to reconsider the initial spectroscopic data which characterized the functional groups to see what additional interpretations might be made in the light of the structure of the carbon skeleton.

Whereas the identification of the functional groups in the preliminary characterization involves chemical and spectroscopic properties that are specific for individual groups in isolation, in this later phase of the study we are concerned with those properties that reflect interactions between functional groups, or between a functional group and a spectroscopically clearly identifiable part of a structure such as a methyl group. Although the information on relationships between groups may be considered under the separate headings of spectroscopic and chemical data, the two overlap and the spectroscopic changes consequent upon a chemical reaction are often very informative.

3.2   Spectroscopic Interrelationships

3.2.1 NMR Methods

The chemical shift, multiplicity and integral mode of the NMR spectrum can each be used to give information on the detailed environment of a nucleus and its relationship to its neighbours.

The **chemical shift** of a proton resonance is a reflection of the magnetic environment of that nucleus. Apart from major effects arising from changes in the immediate environment, for example the insertion of a hydroxyl group, more subtle effects can reflect intramolecular interactions. The ^{1}H NMR signal of a methyl group on the carbon skeleton, particularly one attached to a fully substituted carbon atom as in many terpenoids and steroids, makes a useful structural probe. For example, the angular methyl groups of the steroids at C-10 and C-13 (H-19 and H-18, respectively) are susceptible to **transannular interactions** with functional groups at many of the positions on the steroid nucleus (Scheme 3.1). These methyl groups provide characteristic singlets in the ^{1}H NMR spectra of the steroids in the range δ_H 0.70–1.30. The detailed position of the methyl group resonance within this range reflects these interactions. Thus a 1,3-diaxial interaction with a hydroxyl group can produce a significant deshielding of 0.2–0.3 ppm. These contributions to the chemical shift of the methyl group have been analysed for many hundreds of steroids. An assumption is made that the contributions for individual functional groups are approximately additive, provided that the groups are sufficiently separate from one another. Using these tables, known as the Zurcher tables, it is possible to predict the position of a methyl resonance for variously substituted steroids and compare this with the observed value. The susceptibility of the chemical shift to 1,3-diaxial interactions is not restricted to methyl groups. For example, the 3α-proton resonance is significantly affected by a 5α-hydroxyl group.

The steroids have a well-defined rigid carbon skeleton and form valuable models for exploring the stereochemical aspects of spectroscopic methods and chemical reactions.

The nucleus is surrounded by an electron cloud which "shields" it from the applied magnetic field. The more dense the electron cloud, the greater that shielding and the lower the chemical shift (δ). Substituents which decrease the electron density, e.g. those which are electronegative, are deshielding and lead to a higher chemical shift, e.g. CH_3I δ_H 2.2, CH_3F δ_H 4.3.

Although these effects are well established with the steroids, this type of reasoning can be applied to any series of related secondary metabolites and their derivatives.

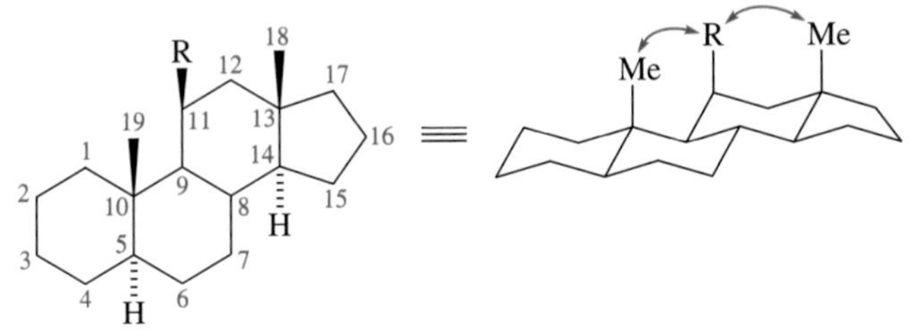

Scheme 3.1 Transannular interactions

The electronic effects of substituents on an aromatic ring are reflected in the chemical shift of the aromatic proton and carbon resonances. Here again there are approximate additivity relationships which may allow assignment of the orientation of specific functional groups to be made.

Relationships, particularly 1,3-diaxial interactions, that affect the chemical shift of key proton resonances may be augmented by the use of aromatic solvents which selectively solvate polar groups. The **aromatic solvent induced chemical shift** differences of resonances between spectra that are determined in deuterochloroform and deuterobenzene or deuteropyridine can be quite revealing. Furthermore, they can help by resolving complex overlapping multiplets. **Shift reagents** such as europium tris(heptafluorodimethyloctanedione) [(Eu(fod)$_3$] have a similar effect. These reagents function by associating with the non-bonding electrons in a substituent, *e.g.* a hydroxyl group. The effect on the chemical shift arises from paramagnetism associated with the unpaired electron on the europium. This through-space effect falls off as the cube of the distance between the nucleus concerned and the europium atom.

In the carbon-13 NMR spectrum, a substituent such as a hydroxyl or carbonyl group has a significant effect on the chemical shift not just of the carbon to which the oxygen is attached but also to that of adjacent carbons. A carbon attached to a carbonyl group may be deshielded by up to 10 ppm compared to the deoxy derivative. Since the nature of the carbon atoms which show these changes (methyl, methylene, methine or quaternary carbon) may be known from the DEPT experiment (see p. 44), this may facilitate the location of the functional group. Furthermore, the hydroxyl group may have a shielding effect of up to 4 ppm on a γ-carbon atom which possesses a *gauche* relationship (see Scheme 3.2). This γ-*gauche* shielding can be useful in both locating a hydroxyl group and in establishing its stereochemistry. These features are illustrated for 6β-hydroxytestosterone (Scheme 3.2, R = OH) compared to testosterone (Scheme 3.2, R = H).

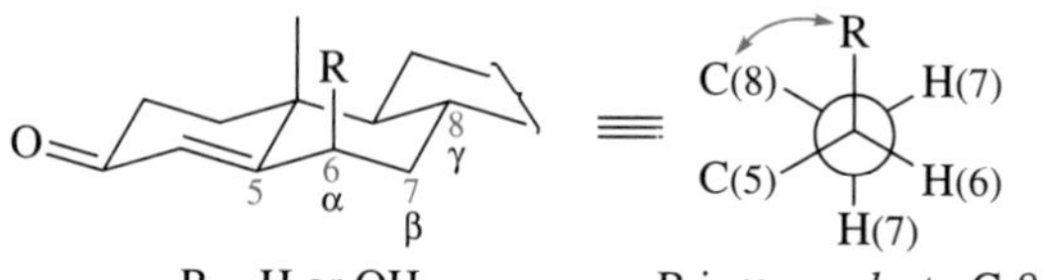

Scheme 3.2 γ-*Gauche* effects

The **nuclear Overhauser effect** (NOE) can play a useful role in locating a hydrogen atom. A proton may relax from its upper spin state by giving out a signal or by the transfer of its magnetization energy to another nucleus. If the internal transfer of energy is prevented by raising the recipient nucleus to its upper spin state by a second irradiation, the original nucleus can no longer relax by this internal mechanism and has to give out more signal. The difference between the spectra determined in the irradiated and unirradiated modes reveals only the signals of those protons that relax via the specific internal transfer that is blocked by the irradiation. An axial methyl group may interact with a number of protons on the carbon skeleton and hence receive energy from them. If the protons of this methyl group are irradiated and transferred to their upper spin state, they can no longer accept this energy and the signals from the protons with which the methyl group interacts are enhanced (see Scheme 3.1). A two-dimensional NOESY spectrum may be plotted linking the various modes of relaxation of hydrogen atoms in a molecule. The coupling patterns which interlink adjacent protons make a valuable method for interrelating functional groups.

The **Karplus equation**, $J = k\cos^2\theta + $ constant, relates the magnitude of the coupling constant J and the dihedral angle θ between two adjacent protons (see Figure 3.1). The original paper emphasizes that the angles obtained from this formula are approximate but are sufficient to distinguish between axial and equatorial substituents (see Table 3.1). An immediate result is that an equatorial proton (*i.e.* an axial functional group) with only small diequatorial and equatorial–axial couplings will produce a significantly narrower signal than an axial proton (*i.e.* an equatorial substituent) with larger diaxial couplings. The influence of adjacent protons on the shape of a signal is sufficiently characteristic in the steroid series to enable, for example, the sites of microbiological hydroxylation to be established from the shape of the CHOH signal.

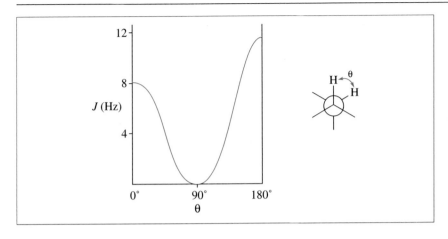

Figure 3.1 The variation of coupling constant (J) with dihedral angle (θ)

Table 3.1 The magnitude of ^1H–^1H coupling constants

Type	J (Hz)	Type	J (Hz)
Geminal coupling			
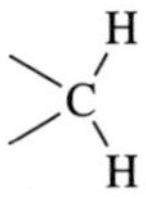	8–20	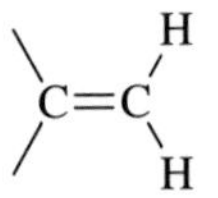	0–3
Vicinal coupling			
Axial–axial	9–12	trans Alkenes	11–19
Axial–equatorial	2–5	ortho Ar–H	7–9
Equatorial–equatorial	0–5	meta Ar–H	2–3
cis Alkenes	5–14	para Ar–H	<1
Long-range coupling			
Allylic	0–2	W-type	0–3

In some rigid systems (*e.g.* Scheme 3.3), **long-range ^4J (four-bond)** couplings can provide useful information. These couplings are significant (*ca.* 2 Hz) when there is a "W" relationship between the hydrogen atoms concerned.

Bridged ring systems occur in a number of natural products. Long-range "W" type couplings are useful in assigning the stereochemistry of hydrogen atoms on these bridges.

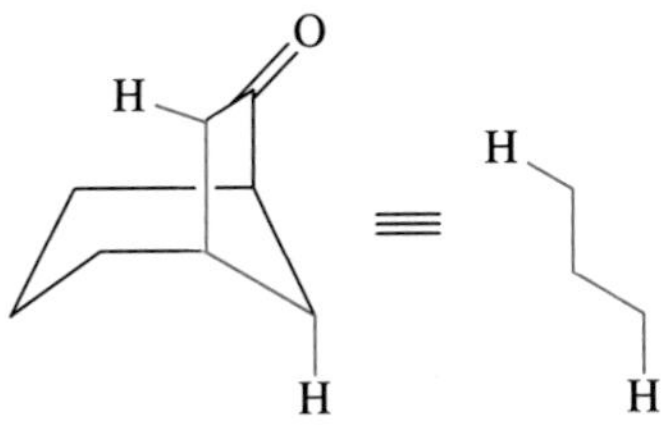

Scheme 3.3 Long-range coupling

The relationships that are suggested by the coupling patterns may be confirmed by spin decoupling experiments.

The magnitudes of the alkene coupling constants are characteristic of the geometry of the alkene and may be used to distinguish *cis*- (J = 5–10 Hz) and *trans*-alkenes (J = 11–19 Hz). Furthermore, the multiplicity may provide information on the number and relationship of the allylic protons. There are also long-range couplings through a double bond to the allylic protons. These also have an angular dependence.

Worked Problem 3.1

Q　The lactone bakayanolide has the two-dimensional structure **1**. The coupling constants between H-6 and the two hydrogen atoms H-7 are 8.6 and 10.9 Hz, respectively, and there is a long-range coupling of 1.8 Hz between H-2 and H-6. There are NOE interactions between H-11, H-6 and H-9. What stereochemical information can be deduced from these data?

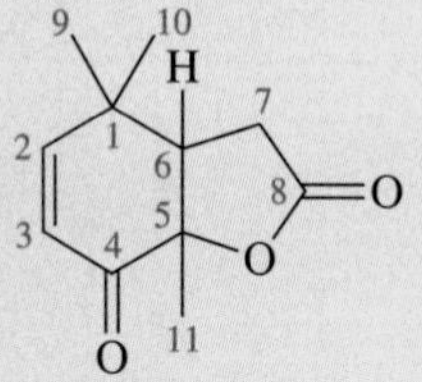

A　The long-range coupling between H-2 and H-6 requires a "W" relationship between these protons and hence an equatorial conformation for H-6. The coupling constants correspond to dihedral angles of *ca.* 0–10° and 150–180° (see Figure 3.1) between H-6 and the hydrogen atoms H-7. The NOE interactions require H-11, H-6 and H-9 to be on the same side of the molecule. This leads to the stereochemistry shown in **2**.

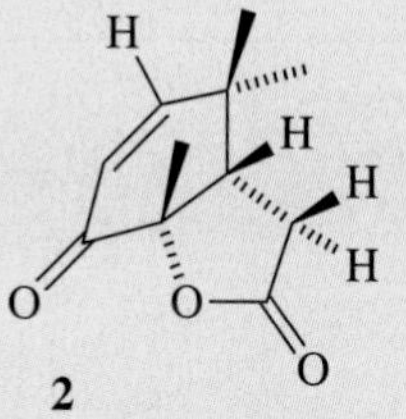

3.2.2　Infrared Spectroscopy

The IR spectrum not only reveals the presence of specific functional groups, such as the hydroxyl and carbonyl groups, but it can also show interactions between them. When a hydroxyl group is involved in **hydrogen bonding**, as in the case of 1,2-glycols, correlations have been established between the length of the hydrogen bond and the position of the hydrogen bonded O–H absorption in the IR spectrum. This was at one time proposed as a method for determining the relative stereochemistry of vicinal glycols, but it has been superseded by more powerful NMR methods. The carbon–oxygen single-bond stretching absorption in the region of *ca.* 1000 cm^{-1} can be used to distinguish between axial and equatorial epimers.

3.3 Chemical Methods

In this section we will describe some general chemical strategies which can reveal useful information on the position, relationships and stereochemistry of a functional group on the underlying carbon skeleton.

3.3.1 Oxidative Strategies

The spectroscopic changes that accompany the oxidation of an alcohol with chromium(VI) oxide to give a ketone, or an aldehyde and then an acid, can be very useful. The IR absorption and ^{13}C NMR signals associated with the new carbonyl group (see Table 2.1), the changes to adjacent carbon resonances, and the changes in the multiplicity of proton resonances are all features which may serve to locate the original hydroxyl group. Base-catalysed deuterium exchange has been used to determine the number of hydrogen atoms on the carbon atoms adjacent to a carbonyl group. For example, one piece of evidence for the location of a ketone at the biosynthetically unusual C-2 position of **ketomanoyl oxide** (**3.1**, Scheme 3.4) came from the incorporation of four deuterium atoms (analysed by mass spectrometry) in a base-catalysed exchange experiment. The C-2 position is unique in this carbon skeleton in being flanked by two methylenes. This information supplemented the dehydrogenation experiments described earlier (see p. 52).

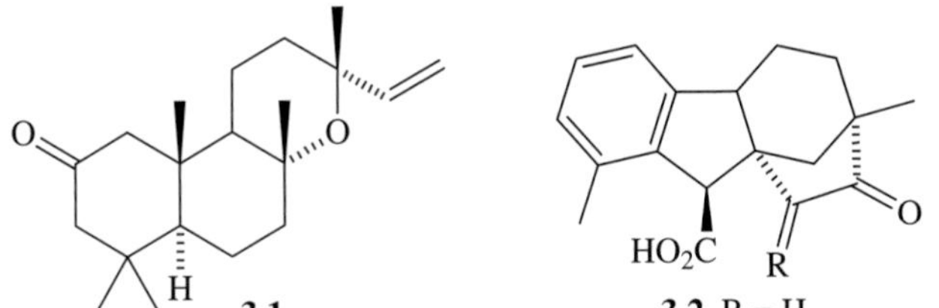

3.1

3.2 R = H$_2$
3.3 R = O

Scheme 3.4   The location of a ketone

Oxidation of the position adjacent to a ketone with selenium dioxide can give an **α-diketone**. The formation of a non-enolizable α-diketone (**3.3**) by oxidation of **gibberic acid** (**3.2**, Scheme 3.4) indicated that the methylene ketone of the five-membered ring was attached to fully-substituted centres, preventing the diketone from enolizing.

The **Baeyer–Villiger** oxidation of ketones to lactones with a peroxy acid has provided a way of opening a cycloalkanone during structural work. For example, it was used to open the seven-membered ring of **apoaromadendrene** (**3.4**, Scheme 3.5) to give **3.5**. This procedure is useful, since it allows a stepwise degradation of a ring via **3.6** and thus the location of substituents, in this case the cyclopropane ring in **3.7**.

Apoaromadendrene was obtained by ozonolysis of aromadendrene, which is a widespread sesquiterpene found in, for example, eucalyptus oil.

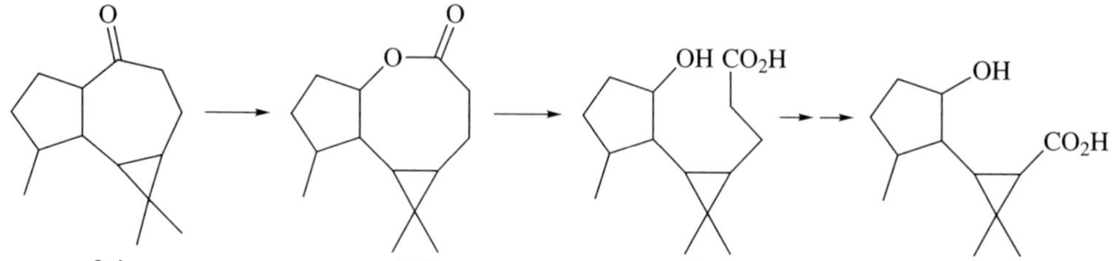

3.4 → 3.5 → 3.6 → 3.7

Scheme 3.5 Degradation of apoaromadendrene

Clavatol is a metabolite of the fungus *Aspergillus clavatus*.

The removal of the acetyl group in the degradation of **clavatol** (**3.8**, Scheme 3.6) with alkaline hydrogen peroxide to give 3-hydroxy-2,6-dimethylbenzoquinone (**3.9**) was used to establish the structure of this fungal metabolite.

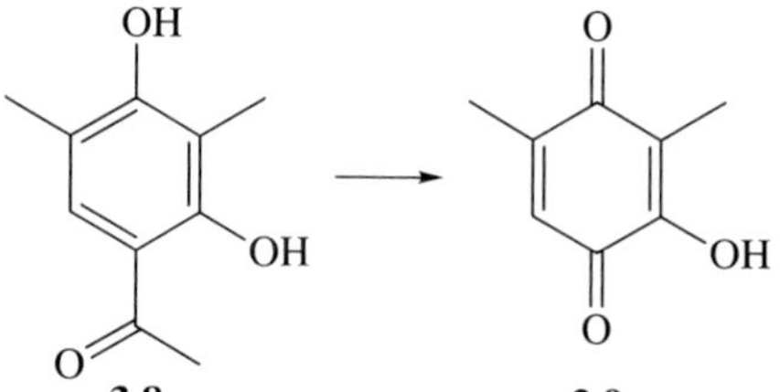

3.8 → 3.9

Scheme 3.6 Oxidation of clavatol

Oxidation reactions can reveal a number of relationships between functional groups. Oxidation of a 1,2-diol with chromium(VI) oxide gives firstly a 1,2-diketone and then a dicarboxylic acid. Pyrolysis of the dicarboxylic acid derived from a cyclohexanone can lead to a cyclopentanone.

Sodium iodate(VII) (or periodic acid) is a specific reagent for the cleavage of a 1,2-diol (*e.g.* **3.10**, Scheme 3.7) to a diketone (**3.11**). Since the reagent forms a cyclic intermediate involving both hydroxyl groups, there are specific stereochemical requirements for this reaction. In particular, a *trans* diaxial diol does not undergo reaction. Cleavage reactions of this type have played an important role in determining the structures of variously modified sugars, particularly those in which some of the hydroxyl groups are methylated. An example of its use came in establishing the position of the free hydroxyl group in the **novioside** derivative **3.12** which was obtained from the antibiotic novobiocin. When the anomeric methoxyl group was present, there was no reaction with sodium iodate(VII). However, when this group was hydrolysed, the product **3.13** reacted with sodium iodate(VII), showing that the free hydroxyl group was adjacent to the anomeric position.

The antibiotic novobiocin is a coumarin glycoside which is obtained from *Streptomyces spheroides* and is used in veterinary medicine.

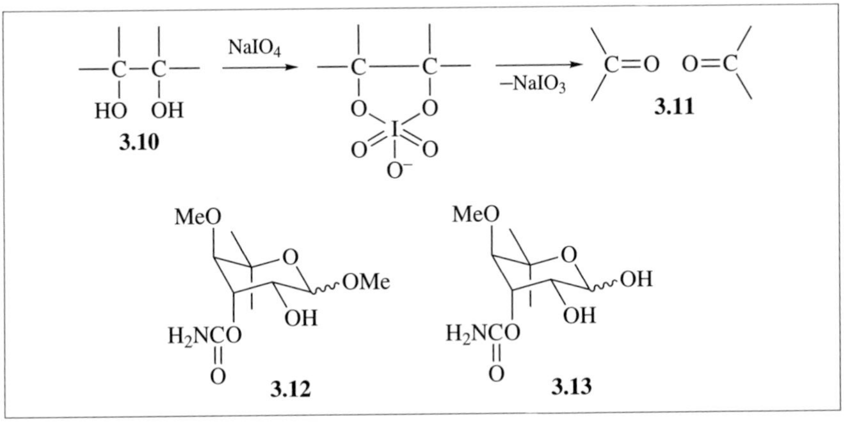

Scheme 3.7 Oxidation of diols with sodium iodate(VII)

The products that are obtained by the oxidation of variously substituted 1,3-diols can reveal their relationship. A β-hydroxy ketone obtained by the oxidation of a secondary/tertiary 1,3-diol can undergo dehydration to an α,β-unsaturated ketone with identifiable spectroscopic characteristics. The formation of a β-hydroxy ketone or a 1,3-diketone permits a number of carbon–carbon bond-cleavage reactions by retro-aldol or retro-Claisen reactions.

The structural feature shown in **3.14** (Scheme 3.8) is quite common amongst the di- and triterpenoids. One example is the plant hormone gibberellin A_{13} (**3.15**). Oxidation with chromium(VI) oxide gave a β-keto acid which readily underwent decarboxylation. The IR spectrum of the product (**3.16**) was typical of a cyclohexanone, but the ^1H NMR spectrum revealed a change in multiplicity (singlet to doublet) for the methyl group. This provided some of the evidence for the structure of the ring A in the original natural product (**3.15**).

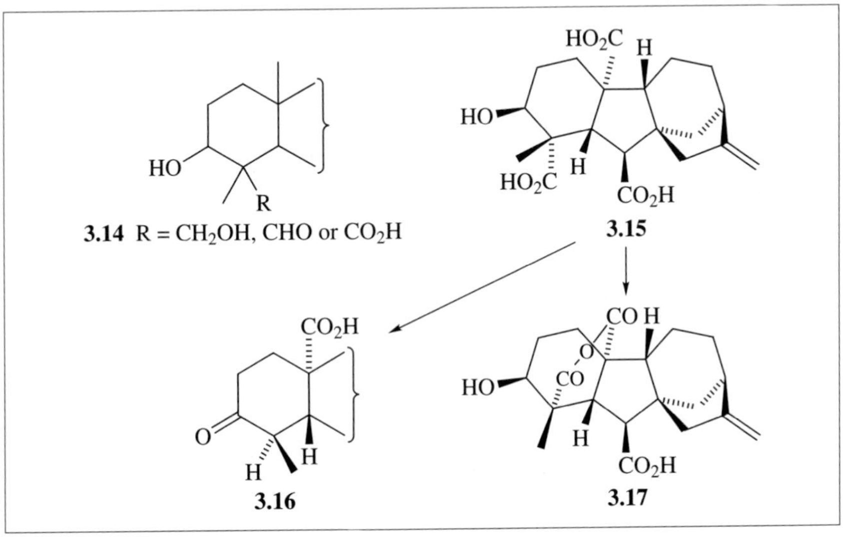

Scheme 3.8 Degradation of gibberellin A_{13}

Salicylic acids (2-hydroxybenzoic acids) are formally the enols of β-keto acids and undergo decarboxylation on heating, revealing a relationship between the phenolic hydroxyl group and the carboxyl group.

Mild oxidation of allylic alcohols to α,β-unsaturated ketones, which may be detected by their spectroscopic characteristics, can show that a hydroxyl group is adjacent to a double bond.

3.3.2 Ozonolysis

The oxidative cleavage of a double bond by ozone has been a valuable degradative method. Many natural products contain an exocyclic methylene ($C=CH_2$). On ozonolysis this affords formaldehyde and a ketone, which can then be the starting point for a further degradation. Thus ozonolysis of gibberellin A_{13} (**3.15**) established the presence of the exocyclic methylene on the five-membered ring.

3.3.3 Ring-forming Reactions

The formation of cyclic structures can serve to show the relationships between functional groups. Geometrical constraints limit the relative positions of participating groups, and thus the formation of a cyclic structure becomes a powerful method for establishing stereochemical relationships.

The formation of **acetonides** from 1,2-diols and acetone (propanone) requires a *cis* relationship between the participating hydroxyl groups. In the pyranose form of **glucose** (**3.18**, Scheme 3.9), this can only be achieved by utilizing the C-1 (anomeric) and C-2 hydroxyl groups. Acetonide formation involving the diequatorial hydroxyl groups at C-3 and C-4 introduces too much angle strain. Diacetone glucose (**3.19**) therefore involves the furanose form of glucose. In contrast, the C-3 epimer of glucose, **galactose** (**3.20**), forms a diacetonide in the pyranose form.

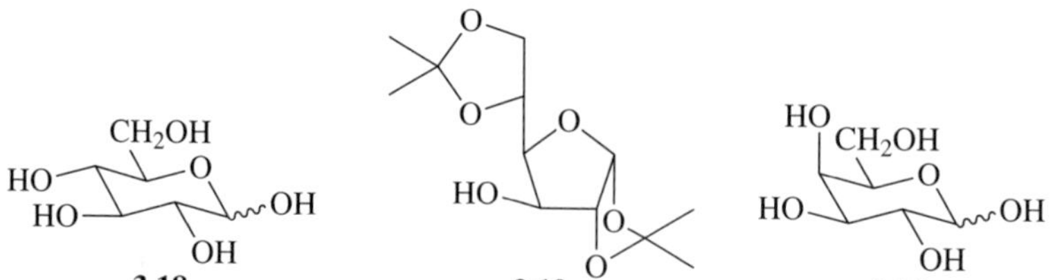

Scheme 3.9   Acetonide formation

Scopine is the basic constituent of the alkaloid scopolamine (hyoscine). Scopolamine is a tropic acid ester which occurs in various *Datura* (thorn apple) and *Hyoscyamus* (henbane) species.

An example of ether formation in structure determination involved the acid-catalysed conversion of the tropane alkaloid scopine (**3.21**) to scopoline (**3.22**) (Scheme 3.10). This defined the relationship between the hydroxyl group and the epoxide in this relative of cocaine.

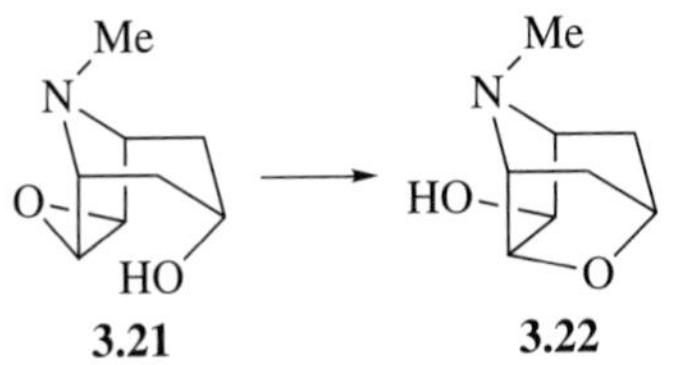

Scheme 3.10 Ether formation in structure determination

The intramolecular esterification of a hydroxyl group by a carboxyl group to form a lactone carries with it stereochemical implications. Similarly, the relationship between two carboxyl groups may be established by the formation of cyclic anhydrides. Thus the C-19 and C-20 carboxyl groups of gibberellin A_{13} (**3.15**) were linked by anhydride formation (see **3.17**).

Worked Problem 3.2

Q The alkaloid platynecine (**3**) is readily converted into the ether anhydroplatynecine (**4**) by thionyl chloride. What stereochemical conclusions can be drawn from this?

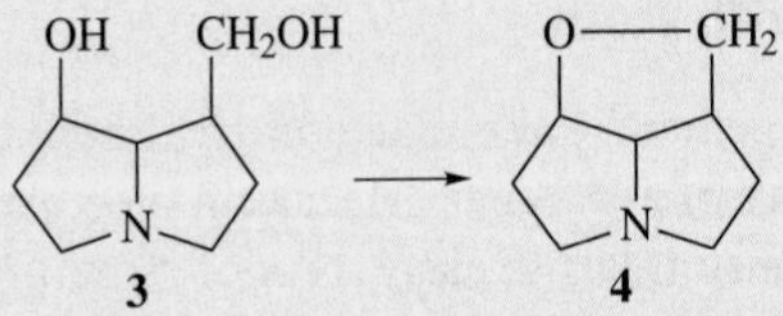

A In order for the ether to form the ring system, the substituents must have the relative orientation as shown in **5**.

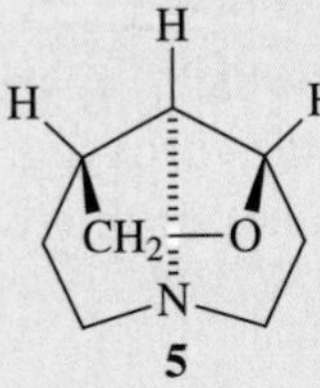

Worked Problem 3.3

Q Ozonolysis of the sesquiterpenoid iresin **6** gives an acetal (**7**) involving the hydroxyl groups and the aldehyde derived from cleavage of the ring B alkene. What stereochemical conclusions can be drawn from this?

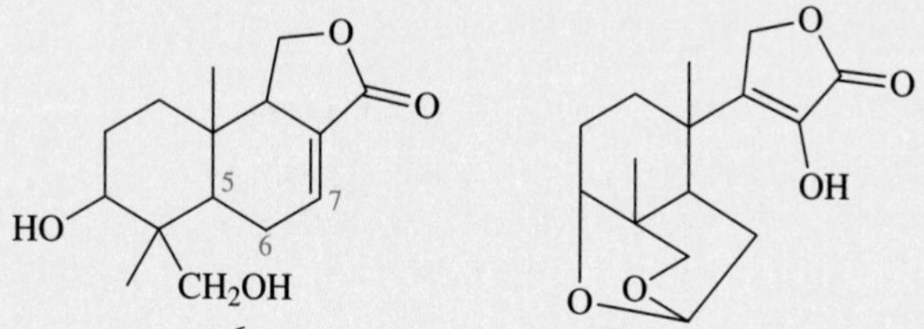

A   Both hydroxyl groups and the 5,6-bond must lie on the same face of the molecule, as shown in **8**.

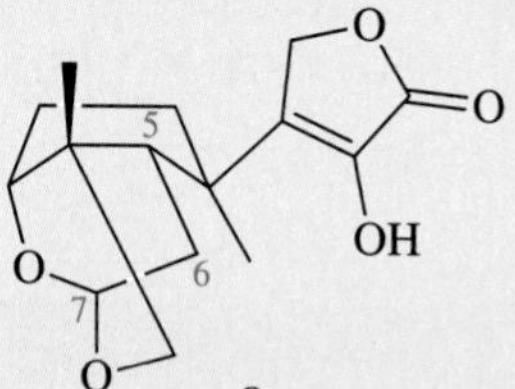

3.3.4 Reaction with Acid

A number of rearrangements take place on treatment of an alkene with acid. Many of these are of a Wagner–Meerwein type. Although they have been a source of confusion in the early days of structure elucidation, the study of these rearrangements has subsequently shed important light on the chemistry of natural products, and it has led to useful interrelationships between groups of compounds.

One example is the acid-catalysed steviol–isosteviol rearrangement (**3.23** to **3.24**, Scheme 3.11) which, apart from providing structural evidence linking the alkene and hydroxyl groups, also served to link the *ent*-kaurene and beyerene families of diterpenoids. Isosteviol (**3.24**) was

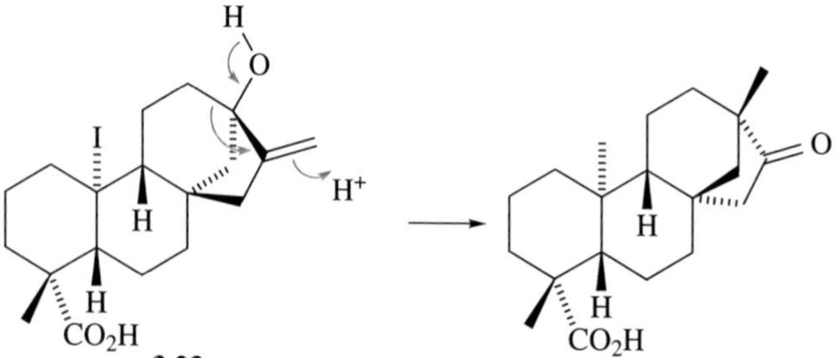

Scheme 3.11 Acid-catalysed rearrangement in structure determination

converted to the beyerene diterpenoid monogynol. A comparable rearrangement played an important role in establishing the structure of gibberellic acid.

3.3.5 Rates of Reaction

The rates of many simple reactions such as **ester formation** and **ester hydrolysis** are influenced by the steric environment of a functional group. The success of the application of these measurements to structure elucidation involves the selection of appropriate models. The differences in rate may enable selectivity to be achieved and a distinction to be made between, for example, two hydroxyl groups. Thus 3β,6α-dihydroxychol-estane (two equatorial hydroxyl groups) readily forms a dicarbonate ester on treatment with ethyl chloroformate. The 6β-epimer with an axial hydroxyl group only forms a 3β-monocarbonate.

One example of the use of the rates of reaction in establishing stereochemistry came from a comparison of the rates of hydrolysis of methyl vinhaticoate (**3.25**, Scheme 3.12) with methyl podocarpate (**3.26**). The former was hydrolysed much more rapidly, and so it was suggested that it was an equatorial ester. Methyl podocarpate was known to possess the axial hindered configuration.

Podocarpic acid is the major constituent of the resin from some New Zealand *Podocarpus* trees. Vinhaticoic acid was obtained from the South American tree *Plathymeria reticulata*. Its structure was of interest because it does not obey the isoprene rule.

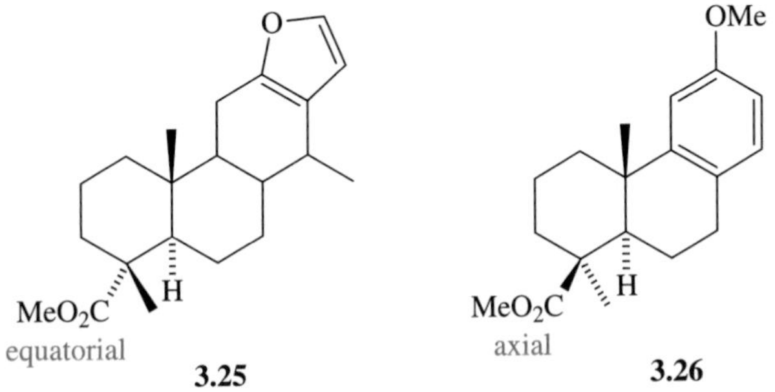

MeO$_2$C H
equatorial
3.25

OMe

MeO$_2$C H
axial
3.26

Scheme 3.12 The rate of hydrolysis of esters

3.4   The Determination of Absolute Stereochemistry

Most natural products possess an asymmetric centre, and normally they occur as only one enantiomer. Where both enantiomers are known they may be obtained from different sources and have different biological properties. For example, the *R*(−)-enantiomer of carvone (**3.27**) tastes of spearmint whilst the *S*(+)-enantiomer tastes of caraway. Although the majority of amino acids occur in the L-form, inversion to a D-amino acid occurs in some biosyntheses, such as that of penicillin. The determination of the absolute stereochemistry of a natural product is an important aspect of structural work.

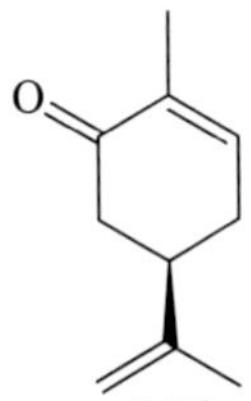

3.27

CHO

H⎯⎯OH

CH$_2$OH

3.28

The possession of chirality by a molecule carries with it the ability to rotate the plane of plane-polarized light. This information is used to characterize the enantiomers of a chiral molecule. The specific rotation $[\alpha]_D^{20}$ of a molecule is defined as the rotation in degrees of the plane of polarization of plane-polarized monochromatic light [measured at the sodium D line (589 nm)] for a solution of one gram of the material in 100 cm^3 of a solvent in a cell of 1 decimetre length at 20 °C, *i.e.* $[\alpha]_D^{20} = 100\alpha/lc$, where α is the observed rotation, c is the concentration in g/100 cm^3 and l is the pathlength (in decimetres). The molecular rotation $[M]_D$ is given by $[\alpha][RMM]$. To give smaller numbers, this is often quoted as $10^{-2}[M]$.

The development of a systematic stereochemical classification of the **monosaccharides**, firstly by Fischer in 1891 and then in 1906 by Rosanoff, led to the convention that the sugars could be divided into two enantiomeric series based on their relationship to either (+)-glyceraldehyde (**3.28**) or its enantiomer. (+)-Glyceraldehyde was arbitrarily assigned the absolute stereochemistry shown in **3.32**. In 1951, Bjivoet determined the absolute stereochemistry of sodium rubidium tartrate by an X-ray method. This had been related to glyceraldehyde, and by a fortuitous coincidence the earlier arbitrary assignment of an absolute stereochemistry to (+)-glyceraldehyde turned out to be correct.

The **absolute stereochemistry** of many of the major groups of natural products was established by a series of careful interrelationships with (+)-glyceraldehyde. A degradation which establishes the absolute stereochemistry at one centre of a molecule is sufficient to determine the absolute stereochemistry of the whole molecule, provided the relative stereochemistry is known. Thus the absolute stereochemistry of a number of simple branched-chain alcohols and carboxylic acids was established by interrelationships in which the chirality of the asymmetric centre was not disturbed. These simple derivatives included (–)-2-methylbutan-1-ol, (+)-2-methylbutanoic acid, (+)-methylbutanedioic acid and (+)-3-methylhexanedioic acid. These compounds then formed useful reference points for the degradation of the monoterpenoids.

3.4.1 Molecular Rotation Differences

The molecular rotation of a molecule may be regarded as the sum of the contribution to the rotation by each of its constituent chiral centres. If two similar asymmetric molecules belonging to the same enantiomeric series are chemically altered in the same way, the change in the molecular rotation will be in the same direction (+ve or –ve) in each case, and will often be of the same order of magnitude. For example, in the sugars it was observed that if the change in the molecular rotation for the conversion of a lactone to the corresponding hydroxy acid was positive, the absolute stereochemistry of the carbon atom bearing the masked hydroxyl group was as shown in **3.29**. A series of molecular rotation differences following acetylation, benzoylation and oxidation of alcohols was used in the steroids and triterpenoids to correlate the stereochemistry of secondary alcohols.

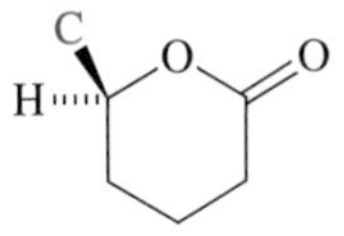

3.29

3.4.2 Asymmetric Induction

The induction of asymmetry through the addition of a Grignard reagent to a carbonyl group forms the basis of the **atrolactic acid** method of determining absolute stereochemistry. Thus when a phenylglyoxalate ester (**3.30**, Scheme 3.13) of a chiral alcohol is subjected to a Grignard reaction and the product (**3.31**) hydrolysed, the chirality of the resultant atrolactic acid can be determined from its rotation. This reflects the chirality around the original alcohol. The validity of this method depends upon the ester preferring the conformation shown in Scheme 3.13.

Atrolactic acid=2-hydroxy-2-phenylpropanoic acid.

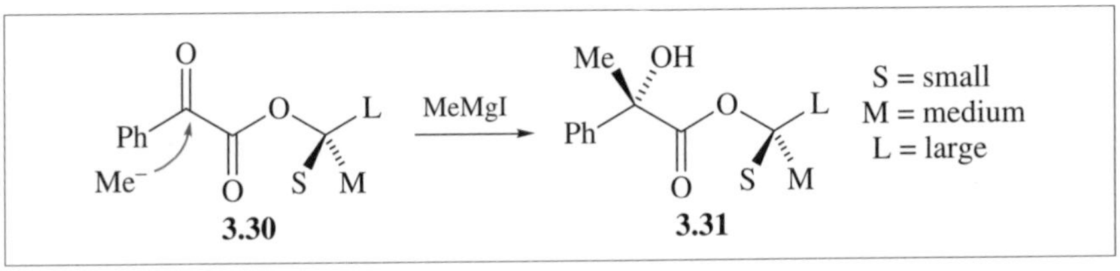

Scheme 3.13 The atrolactic acid method of determining absolute stereochemistry

Horeau's method involves the selective esterification of a chiral secondary alcohol with excess racemic 2-phenylbutanoic anhydride. One enantiomer of the 2-phenylbutyl group reacts preferentially, depending on the steric environment of the secondary alcohol. The optical rotation of the residual 2-phenylbutanoic acid reflects the absolute configuration of the alcohol.

3.4.3 Optical Rotatory Dispersion and Circular Dichroism

The studies that have been described so far relate to the measurement of optical rotation at the sodium D line (589 nm). However, optical rotation varies with wavelength, a phenomenon known as **optical rotatory dispersion**. Stereochemical information can be deduced from this. For some compounds, such as chiral acids, the change of molecular rotation with wavelength is a plain curve in the readily accessible region of the spectrum (Figure 3.2a). For other chiral compounds, particularly those containing a ketone, the curve shows a change of sign (Figure 3.2b), known as the **Cotton effect**, in a region of UV absorption. The sign of the Cotton effect (+ ve or –ve) reflects the chiral environment of the carbonyl group. The contributions of various substituents to the Cotton effect have been evaluated and the results summarized in the **octant rule**. The ketone is viewed in a specific manner (Figure 3.3a) and the space around the ketone is divided into octants by intersecting orthogonal planes. For the majority of ketones the substituents lie in some of the four rear octants, and their contribution may be evaluated as shown in Figure 3.3b.

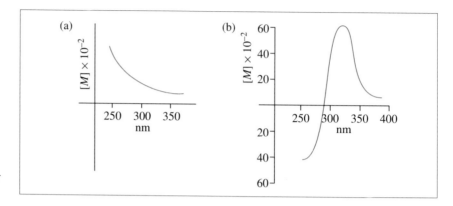

Figure 3.2 Graph of molecular rotation [M] against wavelength

Although there is some theoretical justification for the effect, the correlations have been established empirically, and the value of the method lies in the selection of appropriate model compounds. Other functional groups such as α,β-unsaturated ketones, lactones and dienes also show Cotton effects.

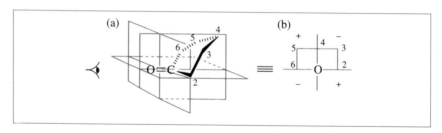

Figure 3.3 The octant rule

Related to this is another phenomenon known as **circular dichroism**. When the difference between the absorption of the left-handed and right-handed components of circularly polarized light is plotted against wavelength, the curve shows a positive or a negative peak in the region of the Cotton effect. The sign of this curve has been correlated with the chiral environment of the absorbing group.

The spatial interactions between two chromophores such as two benzoate esters may also give rise to a Cotton effect. Examination of the dibenzoates of chiral 1,2-diols revealed an intense split π–π^* Cotton effect at 233 and 219 nm. The relationship of the sign of these Cotton effects to the absolute configuration is known as the **exciton chirality** method. The sign of the Cotton effect around 233 nm is in accordance with the chirality: a negative anticlockwise chirality is associated with a negative Cotton effect.

3.4.4 NMR Methods

NMR methods have been developed for the determination of chirality. These use chemical shift differences between esters and amides derived from chiral derivatizing agents, or use chiral shift reagents such as those based on camphor derivatives. Many of these methods have been developed in the context of resolving NMR signals in order to establish the enantiomeric excess of a new chiral centre generated by an asymmetric synthesis.

A useful method has been developed to establish the absolute stereochemistry of secondary alcohols and primary amines using the differences between the chemical shifts of the protons of (*R*)- and (*S*)-2-methoxy-2-phenyl-2-(trifluoromethyl) acetate (**MTPA** acetates, **3.32**; Scheme 3.14). The (*R*)- and (*S*)-*O*-methylmandelate esters of secondary alcohols are also used. The effects arise from the shielding influence of the aromatic ring. The model is shown in **3.32**. The $\Delta\delta$ values $(\delta_S - \delta_R)$ for protons adjacent to the secondary alcohol are diagnostic. They are negative for the protons oriented on the left-hand side of the MTPA plane but for those located on the right side the values are positive. These rules rely on the ester or amide taking up the ideal geometry shown in Figure 3.3a. Although this geometry may be adopted for esters and amides of relatively unhindered equatorial substituents, this is not always the case for more hindered axial substituents.

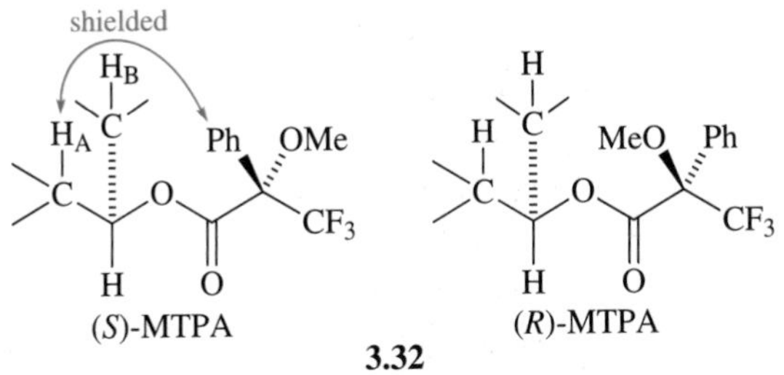

3.32

Scheme 3.14   MTPA method of determining absolute stereochemistry (Mosher's method)

Worked Problem 3.4

Q   The chemical shifts of the neighbouring protons of the 4-(*R*)- and 4-(*S*)-MTPA esters of rosiridin (**9**) are as follows. 4-(*R*)-MTPA ester: H-5 δ_H 2.13, 2.24; H-10 δ_H 1.72. 4-(*S*)-MTPA ester: H-5 δ_H 2.17, 2.28; H-10 δ_H 1.69. What is the absolute stereochemistry of rosiridin?

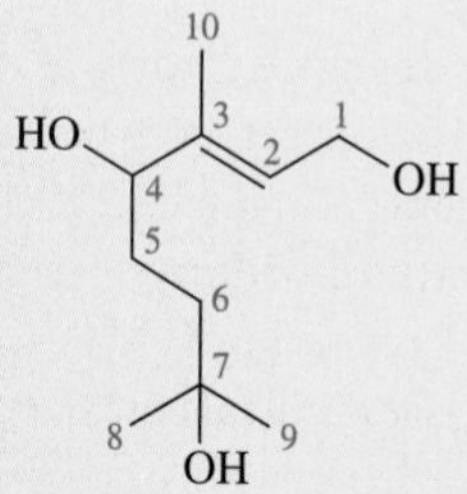

9

A In the model (**3.32**) for the (*S*)-MTPA ester **10**, H_A is shielded relative to the (*R*)-MTPA ester whilst H_B is deshielded. Hence H_A is H-10 and H_B is H-5, *i.e.* the absolute stereochemistry is as shown in **11**. Using the sequence rule the alcohol has the (*R*) configuration.

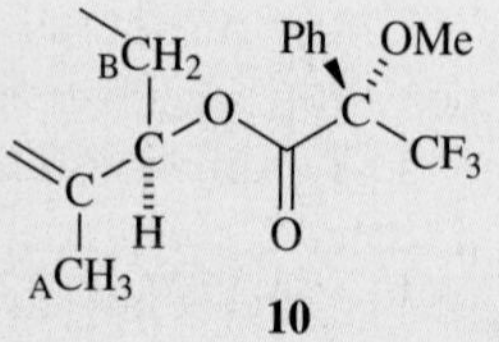

10 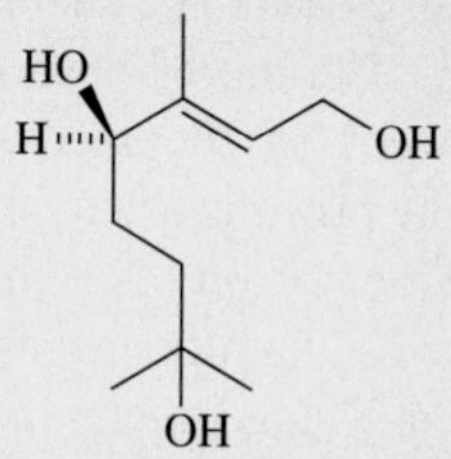   **11**

3.4.5 Crystallographic Methods

Modern X-ray methods utilizing heavy atom derivatives, such as those containing bromine, enabled the absolute stereochemistry of a number of natural products to be established. However, X-ray methods can be used in the absence of a heavy atom if a derivative of the natural product is prepared using a chiral reagent of known absolute stereochemistry. The chiral information associated with the derivative can be extended to the molecule as a whole, and the final X-ray structure gives the absolute stereochemistry of the natural product.

Summary of Key Points

1. ^1H and ^{13}C NMR methods can be used to establish the position and stereochemistry of a functional group on the carbon skeleton.

2.   The magnitude of the coupling constant (J) reflects the dihedral angle between two protons. Nuclear Overhauser effects play a useful role in elucidating stereochemical relationships in a natural product.

3. Oxidation reactions can reveal a number of relationships between functional groups, *e.g.* 1,2-diols.

4. Retro-aldol and retro-Claisen reactions and the decarboxylation of β-keto acids reveal particular 1,3-relationships.

5. Ring-forming reactions *e.g.* ether, acetal and lactone formation, can reveal *cis* relationships between functional groups.

6. Stereochemical correlations, molecular rotation differences, asymmetric induction and spectroscopic methods, such as optical rotatory dispersion (ORD) and circular dichroism (CD) and NMR methods, play a role in establishing the absolute stereochemistry of natural products.

Problems

3.1. A monoterpenoid **A**, $C_{10}H_{18}O_2$, had IR absorption at 3450 cm^{-1} and, in the ^1H NMR spectrum, among other resonances there were signals at δ_H 1.25 (3H, s), 1.65 (3H, s), 3.65 (2H, s) and 5.20 (1H, t). On catalytic hydrogenation, compound **A** took up 1 mole of hydrogen. On oxidation with iodic(VII) acid, compound **A** gave **B**, $C_9H_{14}O$, which had IR absorption at 1700 cm^{-1}. Compound **B** underwent the iodoform reaction to give a carboxylic acid **C**, $C_8H_{12}O_3$, which was dehydrogenated to give 4-methylbenzoic acid. What is the structure of the monoterpenoid **A** and the structures of compounds **B** and **C**?

3.2. The coupling constants between the ring A protons of an iridoid glucoside, 5-deoxypulchelloside, are shown in **D**. What is the relative stereochemistry of the substituents?

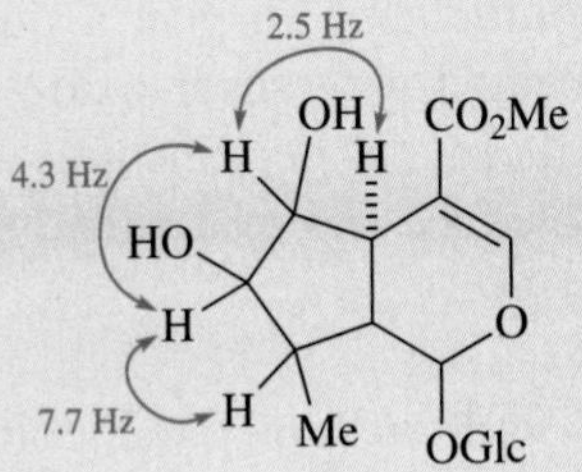

D

3.3. The ethyl *N*-carboxymethyl ester **E** derived from the tropane alkaloids readily cyclizes to form a lactone **F**. The isomer in which the methyl group and carboxymethyl group are interchanged does not form a lactone. What conclusions can be drawn concerning the relative stereochemistry of the hydroxyl groups and the nitrogen bridge?

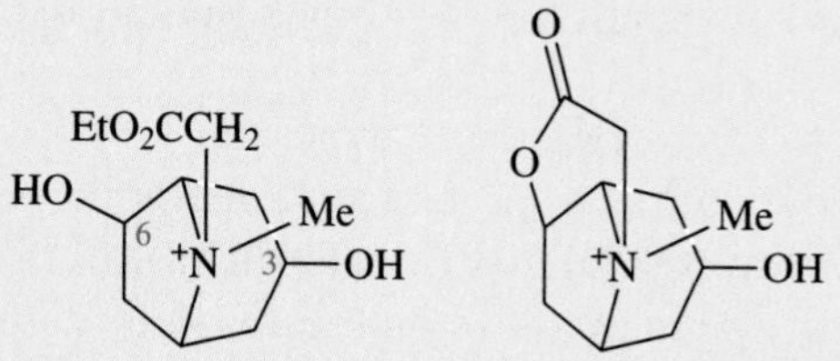

$$E \qquad\qquad F$$

3.4. Alkaloid **G**, $C_8H_{13}NO$, is a tertiary base. It had IR absorption at 3500 cm^{-1} and ^{1}H NMR signals at δ_H 3.0 (5H, m), 3.8 (2H, s) and 5.4 (1H, t, $J = 7$ Hz). It can be oxidized under mild conditions with manganese(IV) dioxide to compound **H**, $C_8H_{11}NO$, which had IR absorption at 1680 cm^{-1} and UV absorption at 229 nm. Compound **H** formed a deep-red dinitrophenylhydrazone. On catalytic hydrogenation, **G** absorbed 1 mole of hydrogen to give **I**. This readily formed a monotoluene-*p*-sulfonate, which on reduction with lithium aluminium hydride gave **J**, $C_8H_{15}N$. When **J** was submitted to three successive Hofmann degradations with a hydrogenation after each stage, the C_8 hydrocarbon 3-methylheptane was obtained. What is the structure of **G**?

3.5. Alkaloid **K** had IR absorption at 1717 cm^{-1} and ^{1}H NMR signals at δ_H 1.5–1.7 (4H, m), 2.4 and 2.8 (each 2H, m), 2.5 (3H, s) and 3.45 (2H, m). It reacts with two moles of benzaldehyde in the presence of base to produce **L**, $C_{22}H_{21}NO$. The alkaloid undergoes vigorous oxidation to produce an acid **M**, $C_8H_{13}NO_4$. The acid **M** reacts with excess iodomethane to give a quaternary ammonium salt. This salt when heated with silver oxide gives an unsaturated acid, which was reduced to heptanedioic acid (pimelic acid), $HO_2C(CH_2)_5CO_2H$. Deduce the structure of the alkaloid **K** and the compounds **L** and **M**.

3.6. On treatment with acetone and toluene-*p*-sulfonic acid, adenosine (**N**) forms an acetonide involving the secondary alcohols of the ribose. The primary alcohol then gives a toluene-*p*-sulfonate on treatment with toluene-*p*-sulfonyl chloride. When the toluene-*p*-sulfonate is treated with sodium iodide, a cyclonucleoside salt (**O**) is formed. What stereochemical information can be deduced from this?

N   **O**

3.7. Antibiotic **P**, $C_7H_{11}NO_3$, had IR absorption at 3120, 2600 and 1650 cm^{-1} and ^{1}H NMR signals at δ_H 1.23 (3H, d, $J = 6.5$ Hz), 3.82 (1H, d, $J = 2.6$ Hz), 5.19 (1H, quintet, $J = 6.5$ Hz), 5.42 (1H, dd, $J = 2.6$ and 4 Hz), 5.83 (1H, t, $J = 6.5$ Hz) and 6.16 (1H, dd, $J = 4$ and 6.5 Hz). Irradiation at δ_H 5.19 collapsed the doublet at δ_H 1.23 to a singlet and converted the signal at 5.83 to a doublet ($J = 6.5$ Hz). Irradiation at δ_H 6.16 also collapsed the signal at 5.83 to a doublet ($J = 6.5$ Hz) and that at 5.42 to a doublet ($J = 2.6$ Hz). Irradiation at δ_H 3.82 collapsed the signal at δ_H 5.42 to a doublet ($J = 4$ Hz). The antibiotic was soluble in base and also gave a hydrochloride. On hydrogenation the antibiotic gave compound **Q**, $C_7H_{13}NO_3$, the ^{1}H NMR spectrum of which lacked the signals at δ_H 5.83 and 6.16. On reaction with diazomethane and then with methyl iodide, compound **Q** gave compound **R**, $C_{11}H_{22}INO_3$. Reduction of this with lithium in liquid ammonia gave a compound $C_7H_{14}O_2$, which was identified by synthesis as 2-(1-hydroxyethyl)-5-methyltetrahydrofuran. What are the structures of the antibiotic **P** and compounds **Q** and **R**?

3.8. A fungal metabolite S, $C_{12}H_{12}O_2$, had IR absorption at 1685 and 1610 cm^{-1} and ^{1}H NMR signals at δ_H 1.7 (3H, s), 3.0 and 3.35 (each 1H, dd, $J = 9$ and 16 Hz), 5.0 (2H, br s), 5.3 (1H, t, $J = 9$ Hz), 6.90 and 7.6 (each 1H, d, $J = 8$ Hz), 7.72 (1H, s) and 9.83 (1H, s). Irradiation of the signal at δ_H 5.3 collapsed the signals at δ_H 3.0 and 3.35 to doublets ($J = 16$ Hz) and irradiation at δ_H 7.72 produced nuclear Overhauser effect enhancements of the signals at δ_H 3.0, 3.35 and 9.83. Reduction of compound **S** with sodium borohydride gave compound **T**, $C_{12}H_{14}O_2$. This compound lacked the IR absorption at 1685 cm^{-1}, but had instead a new absorption at 3620 cm^{-1}. In the ^{1}H NMR spectrum the signal at δ_H 9.83 had been replaced by a two-proton singlet, δ_H 4.6, and an exchangeable resonance at δ_H 2.21. The signals at δ_H 7.6 and 7.72 in **S** had shifted to δ_H 7.2 and 7.26 in **T**. Ozonolysis of compound **T** gave formaldehyde and compound **U**, $C_{11}H_{12}O_3$, which underwent the iodoform reaction. Suggest a structure for compound **S**.

4

Some Examples of Structure Elucidation

Aims

The aim of this chapter is to exemplify the strategies of structure elucidation from the various classes of secondary metabolite. By the end of this chapter you should understand the role of chemical and spectroscopic methods in providing structural information in the context of some specific examples.

4.1 Santonin

The **sesquiterpenoid santonin** (**4.1**, R = H) has attracted considerable interest since it was first isolated from "wormseed" in 1830. It is the best known of a family of sesquiterpenoid lactones which occur in various *Artemisia* species (Compositae) such as *A. maritima*. Santonin has a powerful anthelmintic action and has been used for this purpose in medicine.

Artemisia species are widespread herbs of the Compositae. A number are used in the preparation of various medicines and drinks. *A. absinthium* was used (now banned) in the preparation of absinthe, whilst the herb tarragon is *A. dracunculus*.

An anthelmintic is a compound that is toxic to intestinal worms.

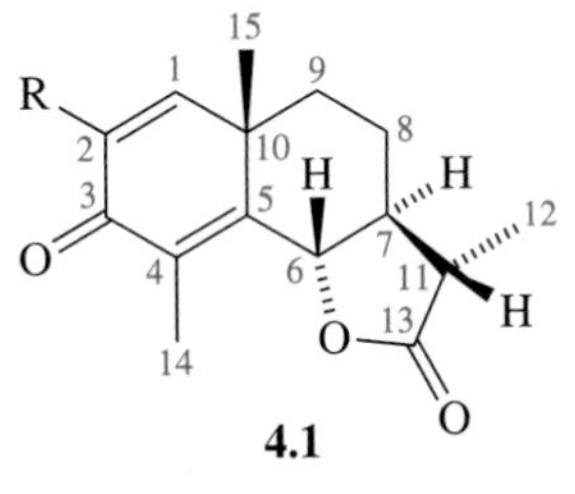

4.1

Much of the earlier work on the chemistry of santonin was carried out by Italian chemists in the latter part of the 19th century and the earlier part of the 20th century. A correct two-dimensional projection of the structure of santonin was proposed in 1929, and the full stereochemistry was finally established by X-ray crystallography in 1962. The studies on the structure of santonin exemplify many of the classical chemical strategies that were used in the era before the dominance of physical methods. These studies were made more complex because santonin undergoes a number of rearrangements. The chemistry of these proved to be very interesting.[1,2]

The molecular formula was established by combustion analysis and relative molecular mass determination as $C_{15}H_{18}O_3$. The **oxygen functions** were characterized as a ketone and a lactone. Santonin formed ketonic derivatives such as an oxime, and the lactone was hydrolysed with alkali to santoninic acid, $C_{15}H_{20}O_4$. The lactone was regenerated from this by treatment with acid. Santonin readily took up two moles of hydrogen to give a mixture of two saturated ketonic lactones, α- and β-tetrahydro-santonins, $C_{15}H_{22}O_3$. Hence there were two C=C double bonds. Since a C_{15} acyclic saturated hydrocarbon would be $C_{15}H_{32}$, there are **seven double bond equivalents** in santonin ($H_{32} - H_{18} = H_{14} = 7$ db equiv). The alkenes account for two of these, the carbonyl groups account for two more, and the lactone ring accounts for one. Two double bond equivalents remained, and hence santonin contained two further rings.

Evidence for the structure of the carbon skeleton came from **dehydrogenation** and **oxidation** experiments. Reduction of santonin with red phosphorus and hydrogen iodide led to hydrogenolysis of the lactone ring and the formation of santonous acid, $C_{15}H_{20}O_3$. Dehydrogenation of this with zinc gave 1,4-dimethylnaphthalene (**4.2**, R = H) and 1,4-dimethylnaphthalen-2-ol (**4.2**, R = OH). The formation of the dimethylnaphthalene accounts for 12 of the carbon atoms, although, as we shall see later, one carbon atom has undergone rearrangement. Nevertheless, this established much of the underlying **carbon skeleton** of santonin.

A lactone is an internal ester. A γ-lactone contains a five-membered ring and a δ-lactone a six-membered ring.

This work was carried out before the discovery of more modern reducing agents.

4.2   4.3   4.4

The **stepwise removal** of the oxygen functions also provided some useful structural evidence. Reduction of santonin oxime with zinc and sulfuric acid gave santoninamine, which on treatment with nitrous acid gave an aromatic compound, hyposantonin, $C_{15}H_{18}O_2$. This was subsequently formulated as **4.3**. Vigorous oxidation of hyposantonin with potassium manganate(VII) gave 3,6-dimethylbenzene-1,2-dicarboxylic acid (**4.4**).

When this evidence was taken in conjunction with the formation of 1,4-naphthalen-2-ol (**4.2**, R = OH), it was apparent that the double bonds and the carbonyl group of santonin are on one ring and the lactone ring was attached to the other. The site of attachment of the latter became clear when the lactone of hyposantonin was hydrolysed to the corresponding hydroxy acid. Dehydrogenation of this gave 7-ethyl-1,4-dimethylnaphthalene (**4.5**, R = Me). The isolation of **4.5** led to the suggestion that a phenolic product of hydrogenolysis of santonin, **santonous acid**, possessed the structure **4.6**. This was confirmed by synthesis.

4.5   **4.6**

A number of structures based on this skeleton were proposed for santonin. However, in 1929, Clemo, Haworth and Walton, in the light of the **isoprene rule** that had been proposed by Ruzicka, suggested that santonin might be a regular sesquiterpenoid containing a carbon skeleton which was constructed from three isoprene units. It was suggested that the formation of a number of santonin degradation products involved the **rearrangement** of a methyl group from the quaternary angular position (C-10) to C-1. Evidence for this was obtained by the reduction of tetrahydrosantonin to deoxytetrahydrosantonin. This compound lacked the dienone which provided the source of the rearrangement. Dehydrogenation of this reduction product gave a naphthalene which lacked the 4-methyl group, 7-ethyl-1-methylnaphthalene (**4.5**, R = H).

Two structural features remained to be established. These were the point of attachment of the lactone ring and the position of the angular methyl group. The point of attachment of the lactone ring was established through the synthesis of desmotroposantonin (**4.3**, R = OH).

The location of the angular methyl group followed from the stepwise oxidative degradation of the dienone of santonin *via* its epoxide to give heptane-2,3,6-tricarboxylic acid (**4.7**), which was synthesized.

The isoprene rule states that the carbon skeleton of terpenoid compounds may be divided into isoprene units linked in a head-to-tail manner (see Chapter 1).

4.7   **4.8**

The eudesmane series of sesquiterpenes have the skeleton:

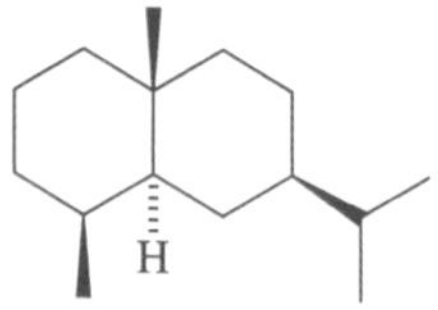

The cadinane and guaiane skeletons are also quite common:

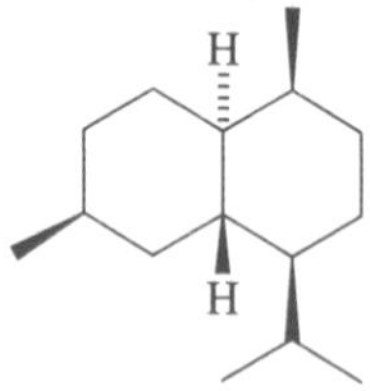

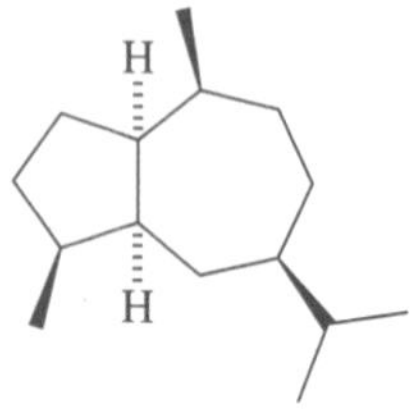

The absolute stereochemistry of β-cyperone had in turn been linked to that of the steroids.

The rearrangement of the C-10 methyl group is a 1,2-shift:

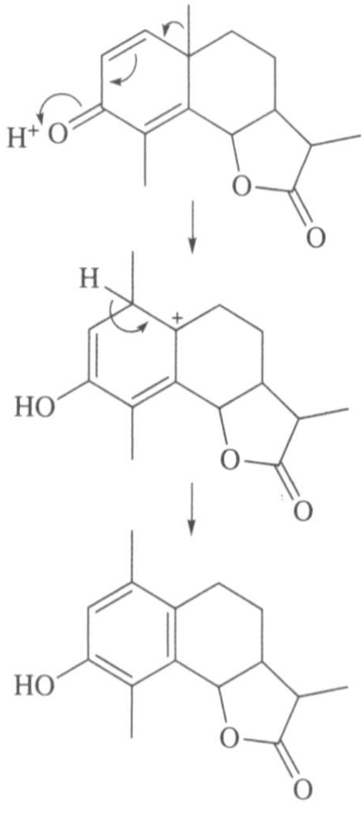

There are four asymmetric centres in santonin, C-6, C-7, C-10 and C-11. Because of the complexity of the rearrangement reactions which santonin undergoes, the final elucidation of the **stereochemistry** took a further 30 years.

Many of the stereochemical arguments were concerned with the stereochemistry of the lactone ring of santonin. When ring A is converted to an aromatic ring, epimerization of the lactone ring can take place by an acid-catalysed mechanism that involves the formation of a planar benzylic carbocation at C-6. Other centres (C-7 and C-11) may be epimerized by elimination–rehydration sequences and by enolization of the lactone carbonyl. A rationalization of the formation of a series of compounds led to the suggestion that santonin possessed a *trans* lactone ring. The controversy regarding the stereochemistry of santonin was ultimately resolved by an X-ray crystal structure of 2-bromo-α-santonin (**4.1**, R = Br).

The **absolute stereochemistry** of santonin was established by linking it to other members of the eudesmane series of sesquiterpenes, and in particular to β-cyperone (**4.8**). Molecular rotation differences and optical rotatory dispersion studies showed that these compounds had the same absolute stereochemistry at C-10 as the steroids.

It is helpful to see how much structural information can be obtained from a straightforward examination of the spectroscopic data, *i.e.* what could be deduced from these data were the structural problem to be examined today. The IR spectrum of santonin contains absorptions at 1784, 1656, 1627 and 1610 cm^{-1}, which are typical of a γ-lactone and α,β-unsaturated ketone. The UV spectrum shows intense broad absorption at $\lambda_{max} = 236$ nm, which would confirm the presence of an α,β-unsaturated ketone. The ^{13}C NMR spectrum shows 15 signals, which from the DEPT spectrum correspond to three methyl groups, two methylenes, five methines and five carbons without an attached hydrogen. Two of the latter (δ_C 186.7 and 178.0) are present as carbonyl groups, and two (δ_C 151.4 and 129.1) are present as alkene resonances. Two of the methine signals are alkene resonances (δ_C 155.3 and 126.3) and one at δ_C 81.6 is a signal for a carbon bearing oxygen. The ^{1}H NMR spectrum contained three methyl group resonances [δ_H 1.22 (doublet, $J = 6.9$ Hz), 1.28 (singlet) and 2.08 (singlet)], corresponding to a CH−CH$_3$, a C−CH$_3$ and a C=C−CH$_3$, respectively. There were two alkene hydrogen signals (δ_H 6.19 and 6.65) which were coupled together ($J = 9.9$ Hz). The chemical shifts of these signals corresponded to those of an α,β-unsaturated ketone. There was a CH−O signal (δ_H 4.75) which appeared as a doublet ($J = 11.0$ Hz), and a doublet of quartets (δ_H 2.37) assigned to the CH−CH$_3$ group. Spin decoupling experiments showed that both the signal at δ_H 4.75 and that at δ_H 2.37 were coupled to the same resonance at δ_H 1.75.

This information provides evidence for the presence of the fragments **4.9**, **4.10** and **4.11** and from the magnitude of the coupling constants suggests that H-6 and H-7 are *trans* to each other.

4.9   **4.10**   **4.11**

If this problem were being studied today, the next steps in the structure elucidation would then be to establish connectivity between these fragments. A two-dimensional $^1H-^{13}C$ COSY experiment would confirm the interrelationship of the ^1H and ^{13}C signals. Since a number of the fragments terminate in fully substituted carbon atoms, an appropriate NMR sequence might be a heteronuclear multiple bond correlation (HMBC) experiment to link these fragments.

4.2   Griseofulvin

Griseofulvin (**4.12**, Scheme 4.1) was first described in 1939 as a metabolite of the fungus *Penicillium griseofulvum*. It is an important polyketide antibiotic which has been used for the treatment of fungal diseases in animals and in man. The structure was elucidated by a combination of chemical and spectroscopic methods. IR and UV spectroscopy played an important part both in the critical evaluation of some early structural proposals and in establishing the correct structure in 1952.[3]

Analytical and relative molecular mass determinations showed that griseofulvin was $C_{17}H_{17}ClO_6$. The preliminary **characterization** of griseofulvin revealed the presence of three methoxyl groups, and the formation of a 2,4-dinitrophenylhydrazone derivative showed that it contained a reactive carbonyl group. Microhydrogenation established the presence of an alkene. One of the methoxyl groups in griseofulvin was hydrolysed under both acidic and alkaline conditions. This hydrolysis gave a weakly acidic product known as griseofulvic acid, $C_{16}H_{15}ClO_6$, which behaved as a 1,3-diketone, indicating that griseofulvin was a β-methoxy-α,β-unsaturated ketone. Griseofulvic acid was subsequently shown to have the structure **4.13** (Scheme 4.1). Catalytic reduction of the hydrolysis product gave a tetrahydro derivative, $C_{16}H_{19}ClO_6$. This compound was a diol and it still retained an unreactive carbonyl group. The UV spectrum showed that this ketone was conjugated with an

Griseofulvin was used as Fulcin for the treatment of fungal infections of the skin, e.g. ringworm.

An sp^2 C–Cl bond is much more difficult to cleave under nucleophilic conditions than an sp^3 C–Cl bond.

aromatic ring. Thus five of the oxygen atoms were accounted for in three methoxyl groups and two carbonyl groups. Since the IR spectrum of griseofulvin did not contain hydroxyl absorption, the remaining oxygen atom must be present as an ether. The chlorine atom was inert towards powerful nucleophiles, and hence it was an aromatic substituent.

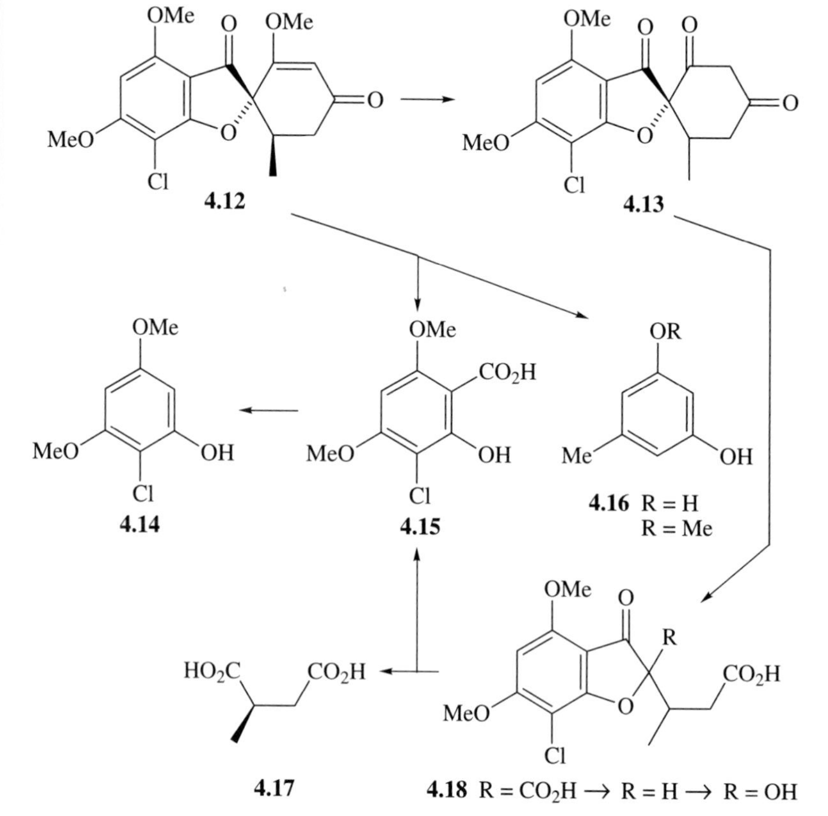

Scheme 4.1 Degradation of griseofulvin

An enol ether is electron rich and sensitive to acidic hydrolysis. The β-position of an α,β-unsaturated ketone is sensitive to nucleophilic addition.

The formula $C_{17}H_{17}ClO_6$ requires **nine double bond equivalents**. One of these was accounted for in the alkene, two in the carbonyl groups and four in the aromatic ring. This left two double bond equivalents unaccounted for, and so apart from the aromatic ring, there were two more rings in griseofulvin.

The **carbon skeleton** and **relative position of the functional groups** in griseofulvin were established by a series of degradations. Typical of a polycarbonyl compound, the most informative of these were a series of **alkaline cleavage** reactions. Fusion with potassium hydroxide gave orcinol, 3,5-dihydroxymethylbenzene (**4.16**, R = H), and sodium methoxide gave orcinol monomethyl ether (**4.16**, R = Me) and the acid **4.15**. The latter were C_8 and C_9 compounds and together they accounted for all 17 carbon atoms of griseofulvin. A characteristic reaction of a 2-hydroxybenzoic acid (salicylic acid) is the loss of carbon dioxide on heating. The C_9 acid gave 2-chloro-3,5-dimethoxyphenol (**4.14**), which was identified by synthesis. This fragment contained two of the methoxyl groups and the third was in the orcinol methyl ether. The isolation of the orcinol methyl ether served to relate the methyl group, the methoxyl group and one of the carbonyl groups of griseofulvin. The base-catalysed cleavage of griseofulvin follows the pathway shown in Scheme 4.2.

The enol of a 1,3-diketone will dissolve in base to give a resonance-stabilized anion.

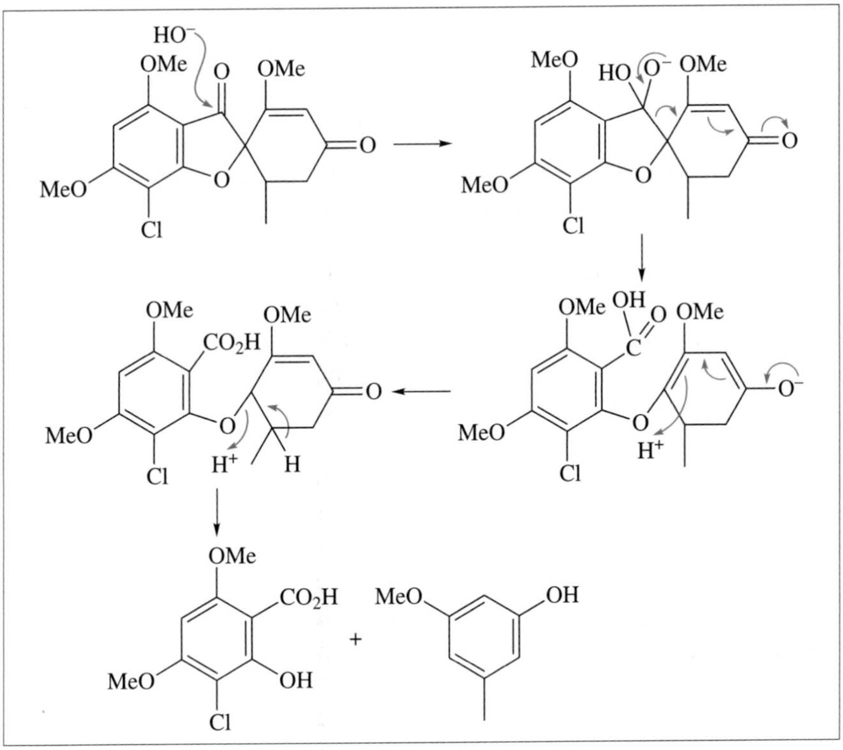

Scheme 4.2 The base-catalysed cleavage of griseofulvin

A stepwise **oxidative degradation** of griseofulvic acid (**4.13**) clarified the relationship of the two six-membered rings. Firstly, oxidation with alkaline hydrogen peroxide gave the dicarboxylic acid (**4.18**, R = CO_2H), which behaved as a β-keto acid and underwent decarboxylation. The product **4.18** (R = H) contained an easily oxidizable C−H adjacent to the carbonyl group. This was hydroxylated with potassium manganate(VII) to give **4.18** (R = OH). Finally, oxidation with iodate(VII) gave the acid **4.15** and (+)-methylsuccinic acid (**4.17**). Since this degradation was performed on griseofulvic acid, it did not distinguish between two alternative enol ether structures **4.12** and **4.19**. These were distinguished by the isolation of 3-methoxy-2,5-toluquinone (**4.20**) from the oxidative degradation of griseofulvin with chromium(VI) oxide.

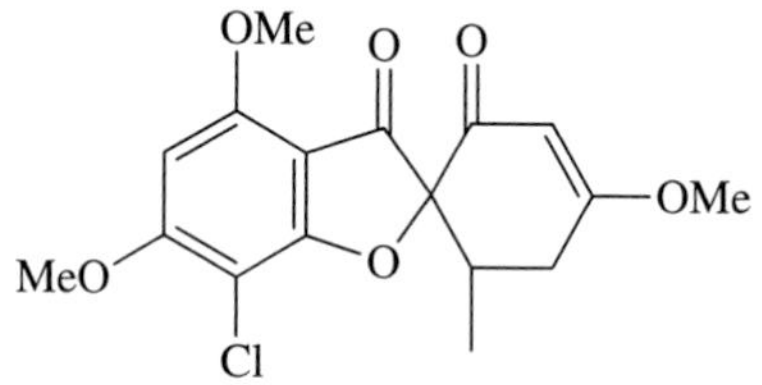

4.19 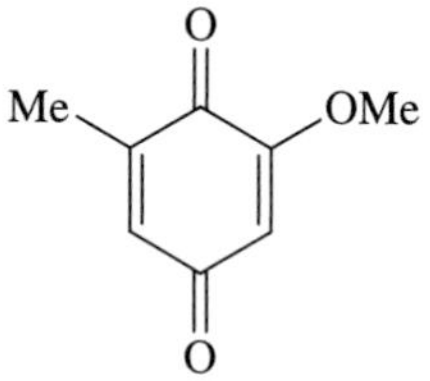   **4.20**

The stereochemistry of griseofulvin was established in a number of studies. The isolation of (+)-methylsuccinic acid (**4.17**) established the **absolute stereochemistry** at one centre. Griseofulvin undergoes epimerization at the spiranic centre. Examination of this equilibrium, and rationalization in terms of the interactions between the secondary methyl group or a hydrogen atom and the acetophenone carbonyl, suggested that the stereochemistry of griseofulvin was **4.12**. X-ray crystallographic analysis of 5-bromogriseofulvin completed the structural work.

A number of the structural features of griseofulvin are apparent in the NMR spectrum, and again it is helpful to review this information to see how the structure might be established today. The ^{1}H NMR spectrum revealed the methyl group doublet (δ_H 0.93, $J = 6.6$ Hz) and three methoxyl signals (δ_H 3.58, 3.94 and 4.00). There were two singlets, one for an alkene (δ_H 5.51) and one for an aromatic proton at δ_H 6.11. There were three other aliphatic proton resonances. A double doublet at δ_H 2.39 ($J = 16.5$ and 4.4 Hz) and δ_H 3.01 ($J = 16.5$ and 13.6) were coupled to each other and to a multiplet at δ_H 2.81. The latter was also coupled to the methyl group doublet at δ_H 0.92. From their chemical shift and multiplicity, this indicated the presence of the system $Me-CH-CH_2-C=O$. The ^{13}C NMR spectrum of griseofulvin has been assigned by using appropriate models, 2-methoxyacetophenone and 2-methoxychlorobenzene for ring A and the ethyl ether of dimedone (**4.21**) for ring C. However, these NMR data on their own would not lead directly to the structure of griseofulvin. Other experiments based on chemical degradation and heteronuclear multiple bond NMR correlations would be required.

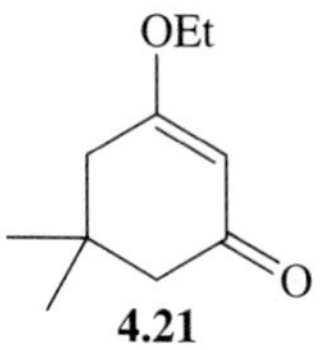

4.21

4.3   Penicillin and Clavulanic Acid

In the summer of 1928, Fleming made the observation that a fungus, *Penicillium notatum*, which was a chance infection of a bacterial culture, was producing a substance that inhibited the growth of various Staphylococci. During the 1930s, several efforts were made to isolate this substance and these came to fruition in 1938 with the work of Florey and Chain. However, the early work was confused because the material that was isolated was a mixture of penicillins. Indeed, owing to the crude nature of the penicillin concentrates which were purified as their barium salts for analysis, the presence of sulfur in the antibiotic was at first missed. Because of the importance of the antibiotic, a major effort to establish the structure was made during the 1939–1945 war.[4]

Various penicillins were isolated, not only from *P. notatum* but also from *P. chrysogenum* and other *Penicillia* and *Aspergilli*. Much of the degradative work was carried out with pent-2-enylpenicillin (penicillin F, $C_{14}H_{20}O_4N_2S$), which was produced on surface culture, and with benzylpenicillin (penicillin G, $C_{16}H_{18}O_4N_2S$), which was produced in submerged culture.

Hydrolysis of the penicillins (**4.22**) with acid produced an α-amino acid, **penicillamine**, $C_5H_{11}O_2NS$ (**4.23**), and a **penaldic acid** (**4.25**). Penicillamine was identified as β,β-dimethylcysteine (**4.23**) by synthesis, and shown to possess the unusual D-amino acid configuration.

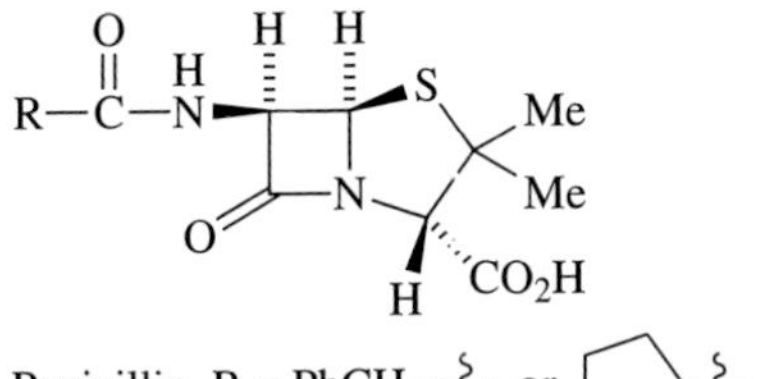

4.22 Penicillin R = PhCH₂–§– or ⬠–§–

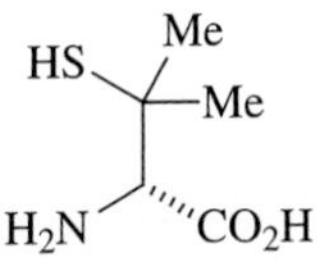

4.23 Penicillamine

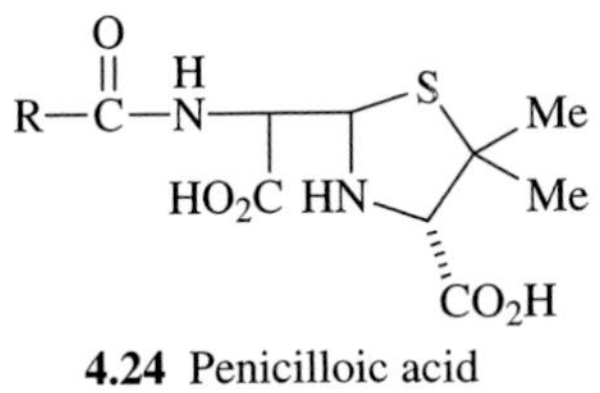

4.24 Penicilloic acid

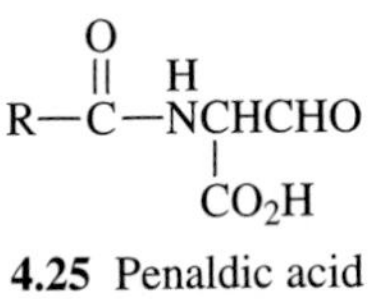

4.25 Penaldic acid

It became apparent that the penicillins differed amongst themselves in the non-penicillamine portion of the molecule, and the view emerged that there existed a penicillin nucleus ($C_9H_{11}O_4N_2S$–R) to which the various substituents (R) were attached. A number of degradations were carried out. A useful sequence centred on the structure of the penaldic acid fragment **4.25**. Alkaline hydrolysis of a penicillin gave a **penicilloic acid** ($C_9H_{13}O_5N_2S-R$, **4.24**), which behaved as a dicarboxylic acid containing a basic amino group. Further alkaline hydrolysis led to the loss of the penicillamine unit **4.23** and the formation of the penaldic acid ($C_4H_4O_4N-R$, **4.25**). Treatment of the penaldic acid with acid led to the loss of carbon dioxide and the formation of a **penilloaldehyde** ($C_3H_4O_2N-R$, **4.27**). This was behaviour that was typical of a β-keto acid.

4.26 Penilloic acid R = PhCH₂-ξ- or ⬠-ξ-

4.27 Penilloaldehyde

The penilloaldehydes were also obtained in another way. Treatment of penicilloic acid (**4.24**) with acid led to the evolution of carbon dioxide and the formation of a monocarboxylic acid, a penilloic acid ($C_8H_{13}O_3N_2S-R$, **4.26**). Hence the penicilloic acids were masked β-keto acids. Removal of the sulfur from the penilloic acid with mercury(II) chloride and hydrolysis of the product gave the penilloaldehydes **4.27**.

The penilloaldehydes gave carbonyl derivatives and also behaved as amides. The significance of the isolation of phenylacetic acid from the degradation of benzylpenicillin then became apparent, and led to the synthesis of the corresponding penilloaldehyde.

Since the penaldic acids behaved as β-keto acids, they were formulated as $RCO-NH-CH(CO_2H)-CHO$.

The question was then posed as to how the penicillamine unit, a cysteine, might be joined to the penaldic acid to give a masked aldehyde. Both amines and thiols add to aldehydes, and hence a structure was proposed for the penicilloic acids which contained a **thiazolidine** ring (see **4.24**). The penilloic acids were the decarboxylation products of this.

The formation of the penicilloic acid from the penicillin involves the addition of water. Although it may at first sight seem surprising, it took some time for this to be seen as the hydrolysis of a cyclic four-membered ring amide. Eventually a β-lactam structure **4.22** for the pencillins was accepted. Part of the reason for this difficulty lay in the structure of this β-lactam ring, which is strained and bent so that there is relatively little amide character in it. The β-lactam is consequently much more reactive than a normal amide, and so were a number of reactions that it was difficult to rationalize when the structure was under investigation.

The thiazolidine ring is a masked aldehyde. When this is opened, it can provide the pathway for the decarboxylation of the penicilloic acids.

Since the discovery of the penicillins, a number of other β-lactam antibiotics, such as the **cephalosporins** (**4.28**), have been discovered.

4.28 Cephalosporin C   **4.29** Clavulanic acid

The enzymatic hydrolysis of the β-lactam ring by β-lactamases is one of the major deactivation routes for the penicillins. The development of resistant strains of bacteria possessing these enzyme systems has become a major problem with these antibiotics. The discovery of **clavulanic acid** (**4.29**), a metabolite of *Streptomyces clavuligerus*, which is an inhibitor of the β-lactamases, was a significant step forward. Its structure determination is typical of a more modern strategy involving spectroscopic methods.[5]

Clavulanic acid (**4.29**) gave a monomethyl ester, the high-resolution mass spectrum of which ($M^+ = 213.0635$) established the molecular formula of the ester as $C_9H_{11}NO_5$. The IR spectrum contained absorption at $\nu_{max} = 3500$ (OH), 1800 (β-lactam) and 1750 cm^{-1} (ester). There was also enhanced alkene absorption at 1695 cm^{-1} which was attributed to an enol ether ($-O-C=C$), thus accounting for the five oxygen atoms. The ^{1}H NMR spectrum contained signals at δ_H 3.05 ($J = 17.5$ and 0.8 Hz), 3.54 ($J = 17.5$ and 2,8 Hz) and 5.72 ($J = 2.8$ and 0.8 Hz), which were assigned to an ABX system $Z-CH_2CHX-Y$. There was also a two-proton signal at δ_H 4.21 which appeared as a doublet ($J = 7$ Hz) coupled to an alkene resonance, δ_H 4.93. The latter also showed a small allylic coupling, $J = 1.2$ Hz, to a signal at δ_H 5.07. Taken together with the IR evidence for the presence of an alcohol and an enol ether, this suggested the presence of partial structure **4.30**.

The ^{13}C NMR spectrum contained nine signals. Those at δ_C 167 and 174 ppm were assigned to the β-lactam and ester carbonyl groups and those at δ_C 152 and 100 (CH) were assigned to the enol ether. There were two CH signals at δ_C 87 and 60 which were assigned to the fragments $O-CH-N$ and $O=C-CH-N$. Two CH_2 signals appeared at δ_C 57 (CH_2OH) and 46 ($O=C-CH_2$) and the methyl ester was at δ_C 53. When these were pieced together they led to the gross structure **4.29** for clavulanic acid. The final structure, including the **absolute stereochemistry**, was elucidated by X-ray crystal structure determinations of the 4-nitro- and 4-bromobenzyl esters.

Clavulanic acid is co-formulated with a penicillin such as amoxycillin in Augmentin, which is used for the treatment of serious infections.

4.30

4.4   Prostaglandins

Prostaglandin preparations such as Dinoprost have been used in medicine for the induction of labour.

The prostaglandins are an important family of hormones that occur in many animal tissues.[6] They possess a wide range of biological activities which include the stimulation of smooth muscle and the regulation of gastric secretions. The prostaglandins and their relatives, the prostacyclins and thromboxanes, also have effects on cardiovascular activity. Their biosynthesis (see Chapter 5) is inhibited by aspirin. The concentrations of the prostaglandins in human seminal plasma are of the order of $10-20$ $\mu g/cm^3$ and they are present in various tissues at levels of 0.5 $\mu g/g$ dry weight. Hence the amounts of material that were available for structural work were very limited. The methods that were used involved micro-scale techniques, including the combination of gas chromatography and mass spectrometry (GC-MS).

The prostaglandins are named as PGA ... PGE, PGF, *etc.*, in the sequence in which they were isolated, with the last letters indicating the nature of the functional groups. Subscript numbers 1, 2, *etc.*, indicate the number of double bonds.

Microanalysis and relative molecular mass determination for prostaglandin E_1 (**4.31**) showed that it had the molecular formula $C_{20}H_{34}O_5$, with four double bond equivalents. Examination of its IR spectrum, together with the formation of a methyl ester and a diacetate, showed that it contained a carboxyl group, a cyclopentanone, two hydroxyl groups and one *trans* double bond. The compound absorbed one mole of hydrogen on catalytic reduction, and on reduction with sodium borohydride the ketone was reduced to give the triol PGF_1, $C_{20}H_{36}O_5$. Three of the double bond equivalents were therefore accounted for by the ketone, the carboxylic acid and the alkene. Hence prostaglandin E_1 had one ring, the cyclopentanone. Prostaglandin E_1 did not have significant UV absorption, and hence the double bond was not conjugated with the cyclopentanone. However, when prostaglandin E_1 was treated with alkali, a compound with strong UV absorption at 278 nm was obtained. The wavelength of this absorption in the UV spectrum indicated that a conjugated dienone (**4.32**) had been formed. This degradation product gave a monoacetoxy-monomethyl ester, *i.e.* it retained one hydroxyl group and the carboxyl group. Ozonolysis of the double bonds in this derivative and oxidative work-up, followed by separation and identification of the products by GC-MS, gave monomethyl suberate (octandioate, **4.33**), succinic acid (butanedioc acid, **4.34**) and 2-acetoxyheptanoic acid (**4.35**). These fragments account for all but one of the carbon atoms of PGE_1. Another derivative (**4.36**) was obtained by treatment of dihydro-PGE_1 with alkali. This compound was an α,β-unsaturated ketone (λ_{max} 237 nm). Ozonolysis of the monoacetoxy-monomethyl ester of this compound gave monomethyl suberate (**4.33**) and 7-acetoxy-4-oxododecanoic acid (**4.37**). The structure of the latter was established by its mass spectroscopic fragmentation pattern.

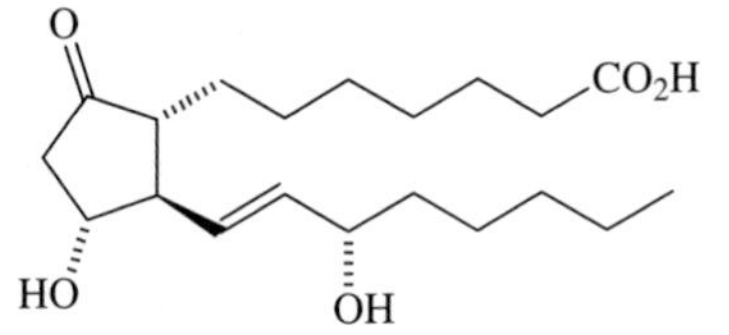

4.31

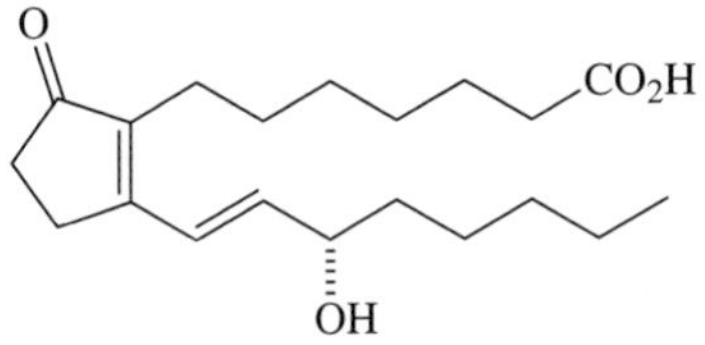

4.32

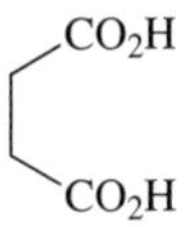

4.34

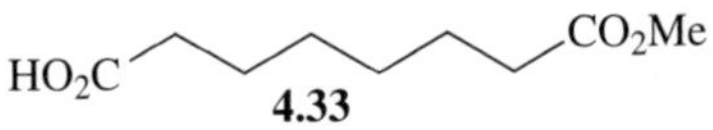

4.33

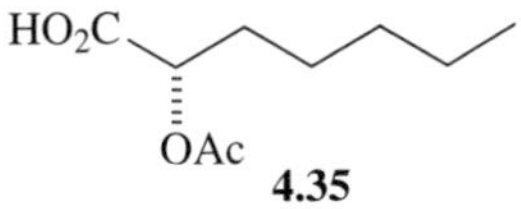

4.35

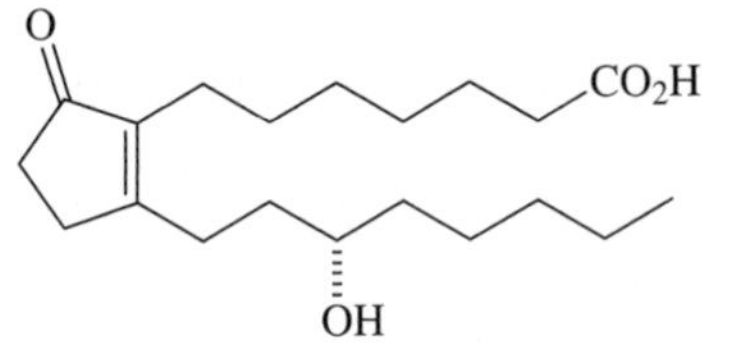

4.36

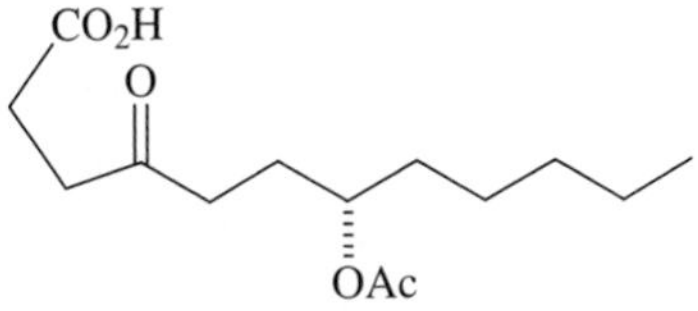

4.37

When this information was brought together, it led to a structural proposal for prostaglandin E_1 (**4.31**). The formation of the dienone may be rationalized in terms of the elimination of the hydroxyl group of a β-hydroxy ketone followed by isomerization of the double bond, as shown in Scheme 4.3.

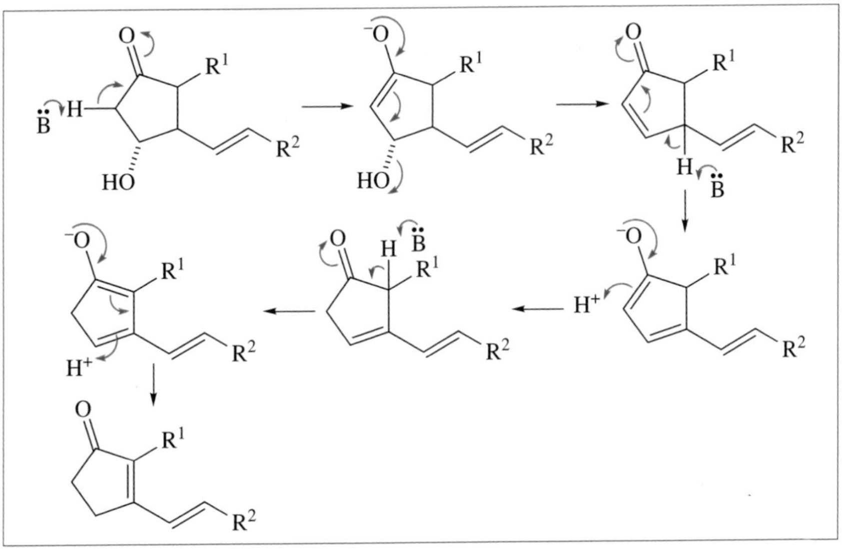

Scheme 4.3 The formation of a dienone in the degradation of prostaglandin E_1

The stereochemistry was established by X-ray crystallographic analysis of the tris(4-bromobenzoate) of the methyl ester of the reduction product PGF_1, and it has subsequently been confirmed by numerous syntheses. The optical activity of the 2-acetoxyheptanoic acid, which was obtained on ozonolysis, originally suggested that the prostaglandins had the "15R" configuration. However, the absolute stereochemistry of the prostaglandins was revised in the light of biosynthetic studies to that shown in **4.31**.

This degradation, carried out on a very limited amount of material, was based on the use of oxidative methods and GC-MS to identify the fragments. GC-MS has become very important in the identification of small amounts of biologically active substances, including many insect pheromones and plant hormones.

4.5   Vitamin C

Scurvy is a major deficiency disease arising from a lack of **vitamin C** (**ascorbic acid**) in the diet. Dietary cures for this disease in terms of fresh fruit, particularly citrus fruits and green vegetables, were recognized in the 18th and 19th centuries. Many efforts were made in the early part of the 20th century to isolate the vitamin. Eventually, Szent-Gyorgi was able to isolate material from the adrenal glands of cattle and from plant sources (paprika), and by 1932 he was able to propose a structure for the water-soluble vitamin C, which became known as ascorbic acid (**4.37**; see Scheme 4.6 below).[7]

The vitamin had analytical data corresponding to $C_6H_8O_6$. Therefore there were three double bond equivalents. It behaved as a weak monobasic acid with the solubility properties of a sugar. Vitamin C had strong reducing properties, and was easily oxidized by mild oxidants such as iodine to dehydroascorbic acid (**4.41**). Dehydroascorbic acid could be easily reduced to ascorbic acid by mild reducing agents such as hydrogen sulfide. This easy reversible redox reactivity was rationalized by the presence of an **ene-diol** in the structure. An ene-diol is tautomeric with an α-ketol (Scheme 4.4).

Scheme 4.4   Oxidation of an α-ketol

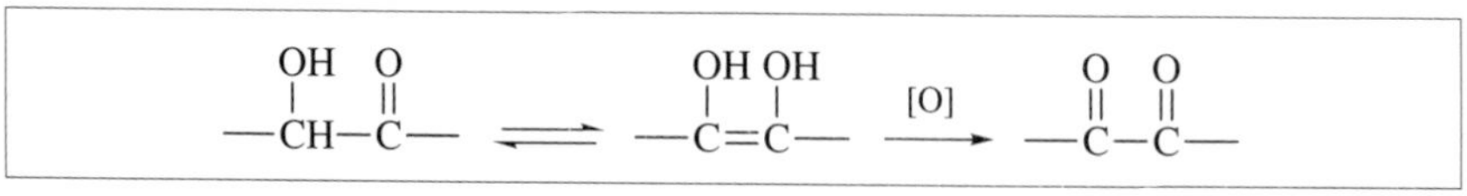

Typical of many sugars possessing an α-ketol, the easily oxidized ascorbic acid reacted with phenylhydrazine to give an osazone, a derivative of a diketone (Scheme 4.3).

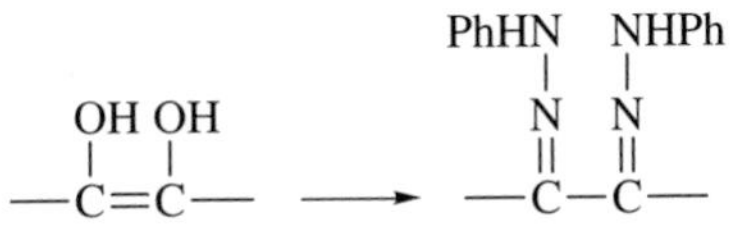

Scheme 4.5 Osazone formation

Treatment of ascorbic acid with diazomethane gave a dimethyl ether (**4.40**; Scheme 4.6), which was no longer acidic. Hydrolysis of this derivative with aqueous sodium hydroxide gave a hydroxy acid without the loss of these methoxyl groups. Hence ascorbic acid contained a lactone ring. There were two more hydroxyl groups which were adjacent to each other, since the dimethyl ether reacted with acetone (propanone) to give an isopropylidene derivative. Hence the six oxygen atoms of ascorbic acid were accounted for by a **lactone ring** and **four hydroxyl** groups.

An isopropylidene derivative is an acetal formed from propanone and a 1,2-diol in the presence of an acid catalyst:

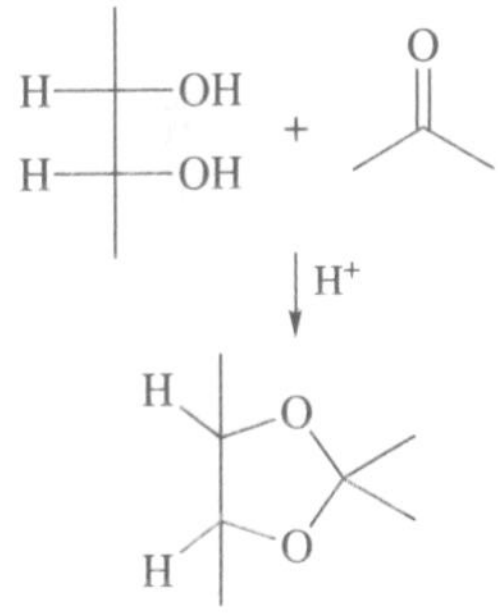

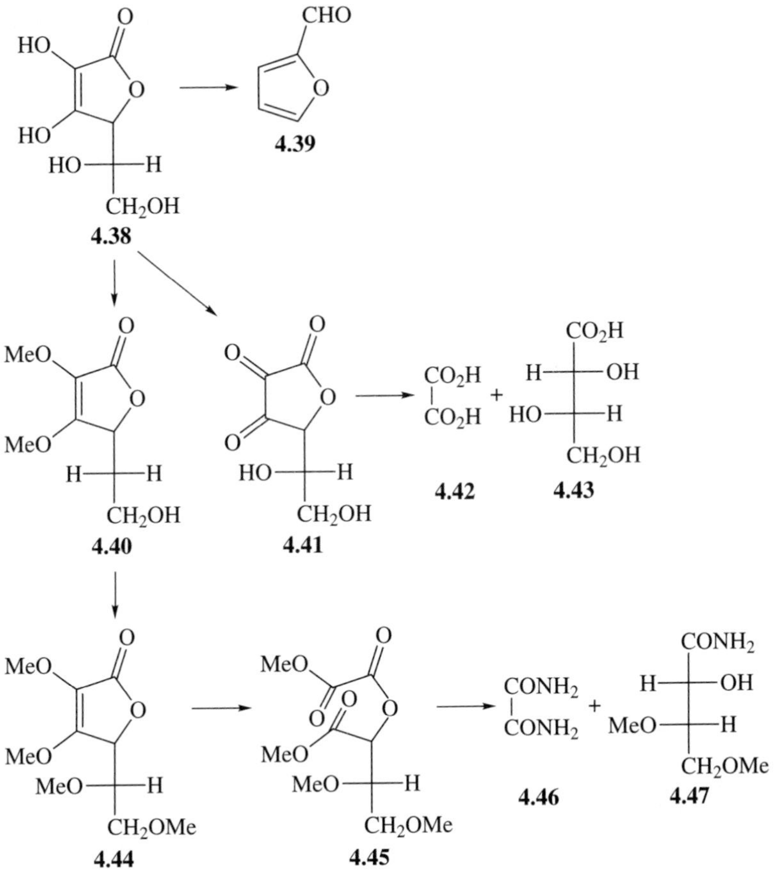

Scheme 4.6 The degradation of vitamin C

Hydrolysis of the lactone gives the enol of a β-keto acid. This decarboxylates under the acidic conditions. Acid-catalysed cyclization and dehydration gives furfural:

$$
\begin{array}{c}
CO_2H \\
H \!-\!\!-\! OH \\
=\!\!O \\
H \!-\!\!-\! OH \\
HO \!-\!\!-\! H \\
CH_2OH
\end{array}
$$

$$
\begin{array}{c}
CH_2OH \\
=\!\!O \\
H \!-\!\!-\! OH \\
HO \!-\!\!-\! H \\
CH_2OH
\end{array}
$$

The location of the methoxyl groups in **4.47** was established by degradation to dimethoxyglyceraldehyde:

$$
\begin{array}{c}
CHO \\
MeO \!-\!\!-\! H \\
CH_2OMe
\end{array}
$$

The nature of the **carbon chain** was revealed by several experiments. Firstly, when ascorbic acid was boiled with hydrochloric acid it gave furfural (**4.39**), and so at least five of the six carbon atoms were in a straight chain. Secondly, further mild oxidation of dehydroascorbic acid (**4.41**) gave the C_2 oxalic acid (ethanedioc acid, **4.42**) and the C_4 L-threonic acid (**4.43**), accounting for all six carbon atoms. The structure and stereochemistry of ascorbic acid (**4.38**) was finally established by a degradation of the methyl ethers. The enolic portion was methylated with diazomethane, and then the remaining hydroxyl groups in **4.40** were methylated with iodomethane and silver oxide to give the tetramethyl ether **4.44**. The double bond of the enol ether was oxidatively cleaved with ozone, and the resultant ester (**4.45**) was hydrolysed with methanolic ammonia. This gave oxamide **4.46** and 3,4-di-*O*-methyl-L-threonamide (**4.47**). The methoxyl groups marked the position of the hydroxyl groups in the original natural product and the free 2-hydroxyl group marked the terminus of the lactone ring. This evidence led to the structure of ascorbic acid, which was substantiated by synthesis in 1935.

The acidic nature of ascorbic acid arises from the enolic system. The electron-withdrawing carbonyl group makes acidic the hydroxyl hydrogen of the enolic β-hydroxyl group. The negative charge of the oxygen anion, which is then formed by ionization of this hydroxyl group, may be delocalized over the carbonyl group of the lactone ring (**4.48** ↔ **4.49**).

4.48   4.49

If the structure of ascorbic acid was being examined today, considerable structural information could be deduced from spectroscopic measurements. The IR spectrum shows four strong hydrogen bonded hydroxyl absorptions at 3210, 3313, 3409 and 3525 cm^{-1}, carbonyl absorption at 1752 cm^{-1} typical of a γ-lactone, and strong alkene absorption at 1672 cm^{-1}. There is significant UV absorption at 243 nm which undergoes a shift to 265 nm in the presence of a trace of alkali, suggesting the presence of an enolic β-hydroxy-α,β-unsaturated carbonyl group. The ^1H NMR spectrum possesses a pair of overlapping double doublets at δ_H 3.44 and 3.45, corresponding to a primary alcohol. These are coupled to a one-proton multiplet at δ_H 3.67, which in turn was coupled to a doublet at δ_H 4.64. Selective decoupling at δ_H 4.64 collapsed the multiplet at δ_H 3.67 to a double doublet, $J = 6.1$ and 7.1 Hz. The multiplicity and chemical shift of these resonances suggests the presence of the system shown in **4.50**.

In agreement with the presence of this fragment, the ^{13}C NMR spectrum contains a methylene at δ_C 61.8 and two methine resonances at 68.6 and 75.9, and there is a lactone carbonyl signal at δ_C 173.0 ppm. The remaining two alkene signals at δ_C 117.5 and 155.2 ppm are typical of an unsaturated ketone. This spectroscopic evidence would lead directly to the structure of ascorbic acid as **4.38**.

$$\begin{array}{c} \text{C}-\!\!\!\diagup \\ \text{HOCH}_2\text{CH(OH)}\overset{|}{\text{C}}\text{H} \\ \diagdown\text{O}-\text{C}-\!\!\! \\ \underset{\text{O}}{\overset{||}{}} \end{array}$$

4.50

Problems

4.1. Apional, $C_{12}H_{12}O_5$, has been isolated from parsley. It had IR absorption at 1684 and 1616 cm^{-1} and ^1H NMR signals at δ_H 9.65 (1H, d, $J = 7.8$ Hz), 7.63 (1H, d, $J = 16.0$ Hz), 6.76 (1H, s), 6.74 (1H, dd, $J = 7.8$ and 16.0 Hz), 6.06 (2H, s), 4.01 and 3.89 (each 3H, s). The ^{13}C NMR spectrum showed signals at δ_C 194.2 (CH), 147.6 (CH), 140.1 (C), 139.5 (C), 138.5 (C), 138.1 (C), 127.7 (CH), 120.0 (CH), 106.7 (C), 102.4 (CH$_2$), 60.4 and 56.8 (CH$_3$). There was a nuclear Overhauser effect enhancement between the resonance at δ_H 4.01 and the signal at δ_H 6.76, and between the resonance at δ_H 3.89 and the signal at δ_H 7.63. What is the structure of apional?

4.2. A natural product **A**, $C_{10}H_{14}O_3$, had IR absorption at 3320, 1630 and 1575 cm^{-1}, UV absorption at 217 and 279 nm and ^1H NMR signals at 1.52 (3H, s), 2.21 (3H, s), 3.46 and 3.81 (each 1H, d, $J = 12$ Hz), 6.57 (1H, d, $J = 8$ Hz), 6.60 (1H, s) and 6.82 (1H, d, $J = 8$ Hz). In addition, it showed signals at δ_H 3.0, 4.6 and 9.0 (each 1H), which disappear on shaking with ^2H_2O. On treatment with sodium iodate(VII), compound **A** gave compound **B**, $C_9H_{10}O_2$, which had IR absorption at 3300, 1680, 1630 and 1575 cm^{-1}, and ^1H NMR signals at δ_H 2.35 (3H, s), 2.59 (3H, s), 6.71 (1H, d, $J = 8$ Hz), 6.78 (1H, s) and 7.61 (1H, d, $J = 8$ Hz). In addition, there was a signal at δ_H 12.3 which disappeared on shaking with ^2H_2O. Compound **B** was identical with the product of the reaction of *m*-cresyl acetate with aluminium trichloride. What are the structures of compounds **A** and **B**?

4.3. Trachelantic acid (**C**), $C_7H_{14}O_4$, on methylation with diazomethane gave a monomethyl ester (**D**), $C_8H_{16}O_4$, which on treatment with acetic anhydride in pyridine gave **E**, $C_{12}H_{20}O_6$.

Reaction of **D** with acetone and anhydrous copper(II) sulfate gave **F**, $C_{11}H_{20}O_4$. Oxidation of trachelantic acid with sodium iodate(VII) gave acetaldehyde (ethanal), 2-methylpropanoic acid (Me$_2$CH-CO$_2$H) and carbon dioxide. What is the structure of trachelantic acid?

4.4.   Terrein, $C_8H_{10}O_3$, had IR absorption at 3390, 3215, 1697, 1636 and 967 cm^{-1}, and UV absorption at 276 nm. It had ^{1}H NMR signals at δ_H 1.89 (3H, d, $J = 7$ Hz), 4.08 and 4.74 (each 1H, d, $J = 2.5$ Hz), 5.95 (1H, s), 6.41 (1H, d, $J = 16$ Hz) and 6.82 (1H, d, $J = 16$ Hz, of q, $J = 7$ Hz). In addition, there were two resonances which disappeared when the solution was shaken with ^{2}H$_2$O. Terrein formed a dinitrophenylhydrazone and this also gave a diacetate. On ozonolysis, terrein gave acetaldehyde whilst the diacetate of terrein gave, in addition, a diacetate of tartaric acid (2,3-dihydroxysuccinic acid, 2,3-dihydroxybutanedioic acid) on oxidative work-up. On hydrogenation, terrein gave a tetrahydro compound, $C_8H_{14}O_3$, lacking the intense UV absorption. Oxidation of this compound with sodium iodate(VII) gave 3-formylhexanoic acid, $C_7H_{12}O_3$. Deduce the structure of terrein.

4.5.   Paeonol, $C_9H_{10}O_3$, on treatment with phenylhydrazine gave a hydrazone and, on acetylation with acetic anhydride, a monoacetate derivative. Paeonol is readily soluble in sodium hydroxide solution but not in sodium carbonate solution. Fusion of paeonol with potassium hydroxide gave 1,3-dihydroxybenzene. Paeonol had IR absorption at 3300 (broad), 1680 and 1570 cm^{-1} and ^{1}H NMR signals at δ_H 2.45 (3H, s), 3.75 (3H, s), 6.41 (1H, s), 6.50 (1H, d, $J = 8$ Hz), 7.55 (1H, d, $J = 8$ Hz) and 11.00 (1H, broad). The signal at 11.00 disappeared when the solution was shaken with ^{2}H$_2$O. A nuclear Overhauser experiment, based on irradiating the signal at δ_H 3.75, produced an enhancement of the signals at δ_H 6.41 and 6.50, whereas irradiation of the signal at δ_H 2.45 produced an enhancement of the signal at δ_H 7.55. Deduce a structure for paeonol.

4.6.   The phytotoxic fungal metabolite pyrenocine C, $C_{11}H_{14}O_4$, had IR absorption at 3410, 1710 and 1640 cm^{-1} and intense UV absorption at 284 nm. It possessed ^{1}H NMR signals at δ_H 1.71 (3H, d, $J = 5.5$ Hz), 2.31 (3H, s), 2.64 (1H, exchanged on shaking with ^{2}H$_2$O), 3.86 (3H, s), 5.15 (1H, d, $J = 7$ Hz), 5.50 (1H, s), 5.60 (1H, d, $J = 15.5$ Hz, of q, $J = 5.5$ Hz) and 5.75 (1H, d, $J = 15.5$ Hz, of d, $J = 7$ Hz). Irradiation of the signal at δ_H 1.71 collapsed the multiplet at δ_H 5.60 to a doublet ($J = 15.5$ Hz), and irradiation of the signal at δ_H 5.15 collapsed the signal at δ_H 5.75 to a doublet ($J = 15.5$ Hz). The signal at δ_H 5.15 received nuclear Overhauser effect enhancements on irradiation of the signals at δ_H 2.31 and 3.86, but the signal at δ_H 5.50 only received an enhancement from irradiation at δ_H 3.86. Suggest a structure for pyrenocine C.

4.7. A natural product, **G**, $C_{11}H_{14}O_5$, had IR absorption at 3400 and 1600 cm^{-1} and UV absorption at 260 nm. It showed ^{1}H NMR signals at δ_H 1.03 (3H, d, $J = 7$ Hz), 2.60 (2H, broad s, exchangeable with ^{2}H$_2$O), 3.74 (1H, quintet, $J = 7$ Hz), 3.90 (3H, s), 4.20 (1H, d, $J = 7$ Hz), 5.95 (2H, s) and 6.55 (2H, s). Irradiation at δ_H 1.03 collapsed the signal at δ_H 3.74 to a doublet ($J = 7$ Hz). Oxidation of **G** with sodium iodate(VII) gave **H**, $C_9H_8O_4$, which showed IR absorption at 1690 and 1600 cm^{-1}. Compound **H** had ^{1}H NMR signals at δ_H 3.90 (3H, s), 5.95 (2H, s), 7.30 (2H, s) and 9.85 (1H, s). There was a nuclear Overhauser effect enhancement of the signal at δ_H 7.30 on irradiation of the signal at 9.85. Deduce a structure for **G**.

References

1. J. L. Simonsen and D. H. R. Barton, *The Terpenes*, Cambridge University Press, Cambridge, 1951, vol. 3, p. 249.
2. J. O. M. Asher and G. A. Sim, *J. Chem. Soc.*, 1965, 6041.
3. J. F. Grove, *Q. Rev. Chem. Soc.*, 1963, **17**, 1.
4. A. H. Cook, *Q. Rev. Chem. Soc.*, 1948, **2**, 203.
5. T. T. Howarth, A. G. Brown and T. J. King, *J. Chem. Soc., Chem. Commun.*, 1976, 266.
6. B. Samuelsson, *Angew. Chem. Int. Ed. Engl.*, 1965, **4**, 410; 1983, **22**, 805.
7. M. B. Davies, J. Austin and D. A. Partridge, *Vitamin C, Its Chemistry and Biochemistry*, Royal Society of Chemistry, Cambridge, 1991.

5
The Biosynthesis of Secondary Metabolites

Aims

The aim of this chapter is to outline the major biosynthetic pathways leading to the polyketides, terpenoids, phenylpropanoids and alkaloids. By the end of this chapter you should be able to:

- Recognize the major building blocks that are used by nature to assemble these secondary metabolites
- Understand the mechanism of formation of the major classes of secondary metabolites
- Rationalize the structures of novel secondary metabolites in terms of a plausible biogenesis

5.1 Introduction

As the structures of natural products were established, chemists began to speculate on their origins. The structural patterns which emerged led to the realization that there were some basic building blocks used by nature to assemble natural products. The structural relationships between natural products suggested common biosynthetic pathways. Although this biogenetic speculation was a useful aid in determining structures, the means to test these theories were not available until isotopic labelling techniques became well established. Biogenetic speculation had nevertheless led to the proposal of the polyketide, terpenoid (isoprenoid) and phenylpropanoid (C_6–C_3) pathways and the role of amino acids in alkaloid biosynthesis.

5.2 Biosynthetic Methodology

A biosynthetic pathway involves the enzyme-catalysed conversion of a series of intermediates ($A \rightarrow B \rightarrow C \rightarrow D$) to a natural product.

Biosynthetic methodology can be divided into **labelling** and **enzymatic** experiments. In the first, a precursor (*e.g.* A) is labelled with a stable or radioactive isotope, and the fate of this precursor is determined when it is incubated with the living system.

In order to establish a sequence of events, the label from A must appear sequentially in B, C and D. Furthermore, for the pathway A → D to be a true biosynthesis and not a degradation and re-synthesis, the label must appear in the metabolites of A specifically at the predicted place and not elsewhere. If the site of the label is randomized, this is an indication of degradation and re-synthesis. With radioisotopes (^{14}C and ^{3}H), this has meant a careful chemical degradation of the biosynthetic product to establish the specificity of labelling. With carbon-13 and deuterium, this specificity may be established much more easily by NMR methods. Mass spectrometry may also be used to establish the incorporation and location of isotopes such as oxygen-18.

The use of double-labelling techniques can establish the structural integrity of a unit. For example, if a precursor is labelled at two or more centres with tritium and carbon-14 and the ratio of the tritium to carbon-14 remains constant between the precursor and the product, it is unlikely that the tritium and carbon-14 labels have parted company during the biosynthesis. Similarly if a ^{13}C–^{13}C coupling is retained in the NMR spectrum, this unit has probably remained intact throughout the biosynthesis.

The success of the ^{13}C–^{13}C coupling experiment in identifying individual acetate units in a chain rests on minimizing inter-acetate unit coupling while observing intra-acetate coupling. This is achieved by diluting the labelled acetate with sufficient unlabelled material to minimize the chances of two adjacent units in any individual chain being derived from enriched material. When the bulk material, made up of many individual chains, is examined, the coupling that is observed arises from within an acetate unit and not between acetate units.

Carbon-13 can also act as a "reporter" label for deuterium in the NMR spectrum. A carbon-13 nucleus attached to or adjacent to a deuterium atom shows a very small but detectable isotope shift compared to the unlabelled species. The **stereochemical** fate of a label, particularly deuterium or tritium, provides useful mechanistic insight into specific steps, for example whether inversion or retention of configuration has occurred.

Interference with the **enzymes** that mediate biosynthetic events provides further information on particular steps. If an enzyme that catalyses a specific step is absent, either in a genetically deficient **mutant** or because the enzyme system has been inhibited or poisoned, the biosynthetic substrates which it normally transforms may accumulate. The product of this particular step then becomes an essential dietary factor for the organism to flourish. Living cells in which biosynthesis

If some small units enter the central metabolic cycles, such as the tricarboxylic acid cycle, the label can become scrambled.

takes place are highly structured. If the cellular structure is disrupted, it may be possible to develop **cell-free enzyme preparations** which just catalyse individual steps. It can be possible to purify the enzyme system itself and obtain it in a crystalline form. The enzyme may then be examined by X-ray crystallography with or without a bound substrate.

Now that it is possible to identify the genetic sequences that code for specific enzymes and to **clone** these systems, this has become a very powerful tool. The transfer and over-expression of genetic information into an easily grown bacterium permits the preparation of quantities of the relevant enzyme systems, and enables individual steps in a biosynthesis to be studied in detail.

In studying biosynthesis, the complementary role of labelling and enzymatic experiments provides information on building blocks, the structures of intermediates together with the sequences, stereochemistry and mechanisms of biosynthetic events.

The over-expression of genetic information involves the incorporation of a sufficient amount of a DNA sequence into the host organism such that the protein for which it codes is produced in a useful quantity.

5.3   The Pathway of Carbon into Biosynthesis

Before considering individual pathways in detail, it is helpful to consider the overall pathway of carbon from carbon dioxide and the photosynthetic formation of sugars. This is shown in Scheme 5.1. In this

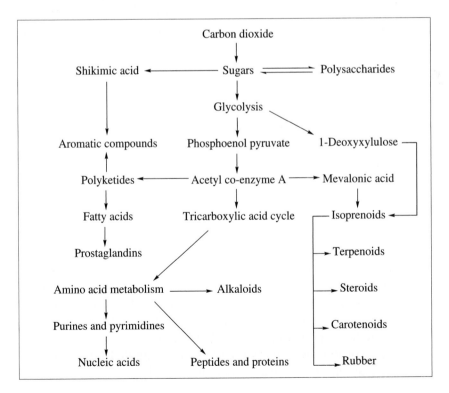

Scheme 5.1 The relationship of natural products

context a distinction may be drawn between primary and secondary metabolic pathways. The former are involved in the storage and release of energy and in the synthesis of essential cellular constituents such as the amino acids and the nucleic acids. The latter involve the biosynthesis of the metabolites which are characteristic of the particular species under investigation. It is with these pathways that we are concerned.

5.4   The Biosynthesis of Polyketides

Some of the earliest experiments on biosynthesis established the role of **acetate** units in the formation of fatty acids and polyketides. The preponderance of fatty acids with even numbers of carbon atoms implied a role for the C_2 unit of acetate in their biosynthesis. Labelling experiments in 1945 showed that a fatty acid, stearic acid, incorporated acetate units and the active acetate unit was identified as **acetyl co-enzyme A (5.1)**. The observation was made that the biosynthesis was dependent on carbon dioxide, biotin (**5.2**) and magnesium ions, but that the carbon dioxide was not incorporated into the final fatty acid. This led to the suggestion that malonyl co-enzyme A (**5.3**) was a precursor, the two carbonyl groups serving to activate the methylene for a condensation with an acetate unit. The difficulty of detecting intermediates of carbon chain length between C_4 and C_{16-18} led to the proposition that a **multi-enzyme complex** existed for the biosynthesis of the fatty acids and polyketides. This is shown in Scheme 5.2.

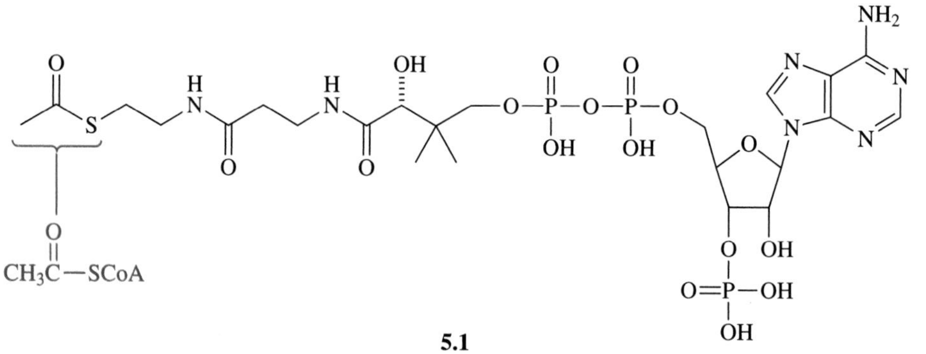

5.1

The C_4 unit, which is formed at the end of the first cycle, is the substrate for the second cycle. The enzyme systems ketosynthase, ketoreductase, dehydratase and enoyl reductase may either be discrete catalytic regions (domains) in one multi-enzyme complex (polyketide synthase, PKS) or a series of single enzymes that are loosely associated.

A feature of fatty acid biosynthesis is that saturated fatty acids are produced first and then dehydrogenation takes place to generate the *cis*

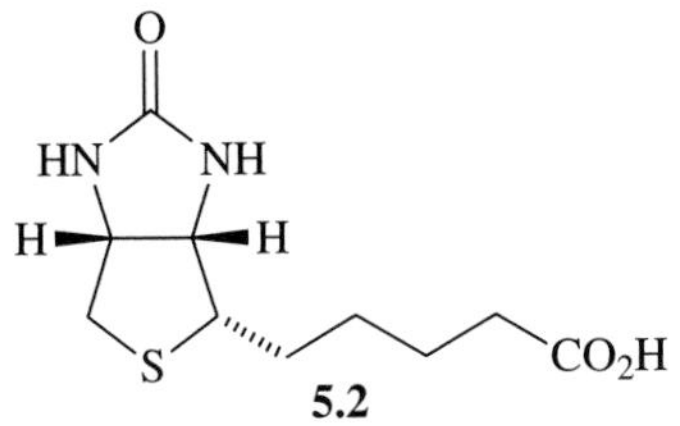
O
HN
NH
H
H
S
CO2H
5.2

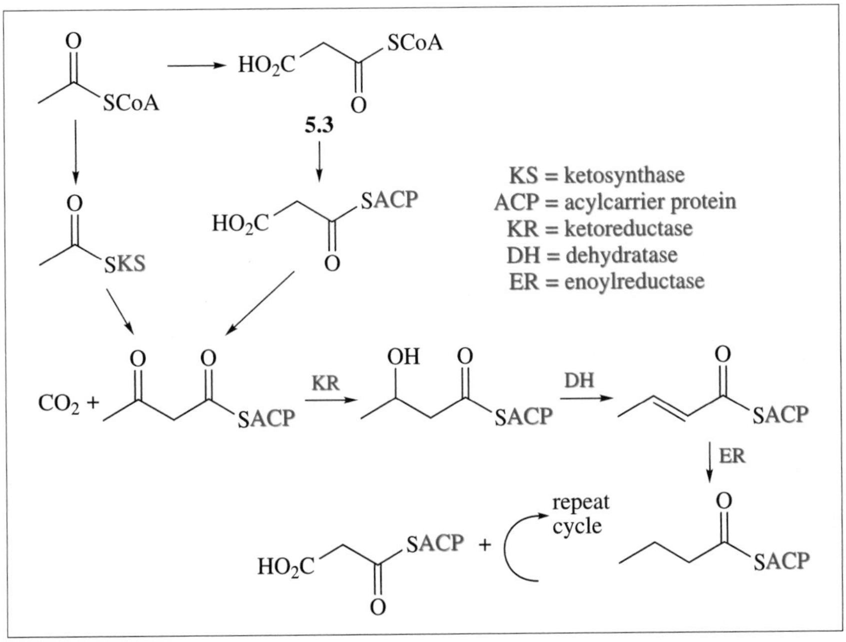
SCoA
HO2C
SCoA
5.3
SKS
SACP
KS = ketosynthase
ACP = acylcarrier protein
KR = ketoreductase
DH = dehydratase
ER = enoylreductase
CO2 +
KR
DH
ER
OH
repeat
cycle

Scheme 5.2 The biosynthesis of fatty acids

double bonds typical of unsaturated fatty acids such as oleic acid (*cis*-octadec-9-enoic acid). In the case of polyunsaturated fatty acids, these double bonds are not normally conjugated. One such unsaturated acid, **arachidonic acid** (**5.4**) undergoes an oxidative cyclization (Scheme 5.3) that is catalysed by **cyclo-oxygenase** to form a family of hormones known as the **prostaglandins** (see Chapter 4). This particular enzymatic step is inhibited by aspirin.

When the reductive steps in fatty acid biosynthesis do not take place, polycarbonyl compounds are formed which are the substrates for various cyclases that lead to aromatic compounds. The formation of **aromatic compounds**, particularly fungal metabolites, from acetate units has been established by many radio- and stable isotope studies. Indeed, the formation of aromatic compounds by the cyclization of polycarbonyl compounds had been suggested as the result of model experiments by

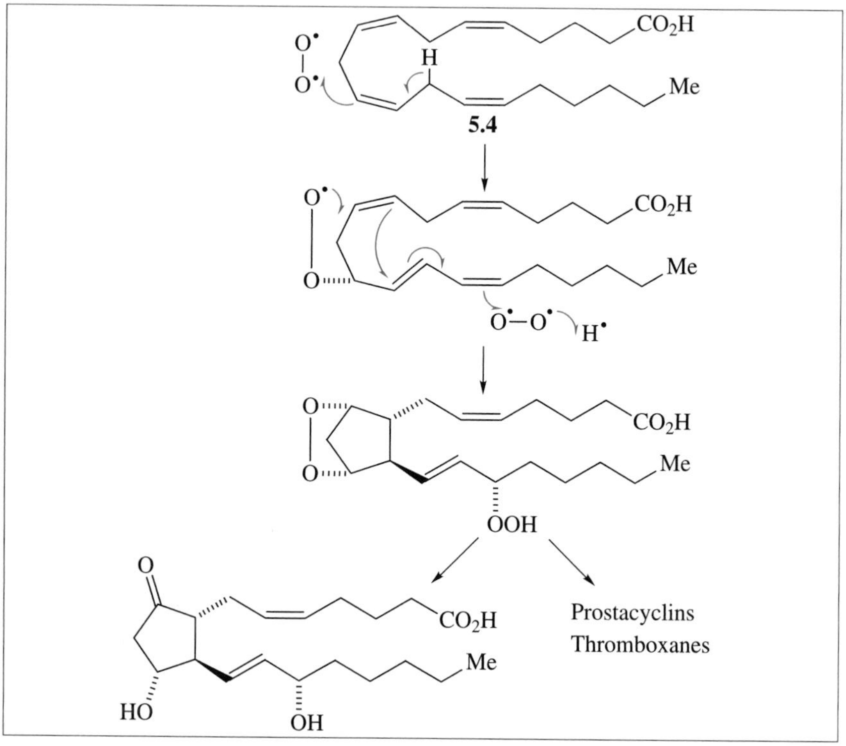

Scheme 5.3   Prostaglandin biosynthesis

Collie in 1906. Experimental work by Birch in 1953 using ^{14}C-labelled acetate established its role in the biosynthesis of 6-methylsalicylic acid (**5.5**; see Scheme 5.4). Subsequently, 6-methylsalicylic acid synthase has been examined in detail, and the sequence and stereochemistry of the steps in the biosynthesis have been established.

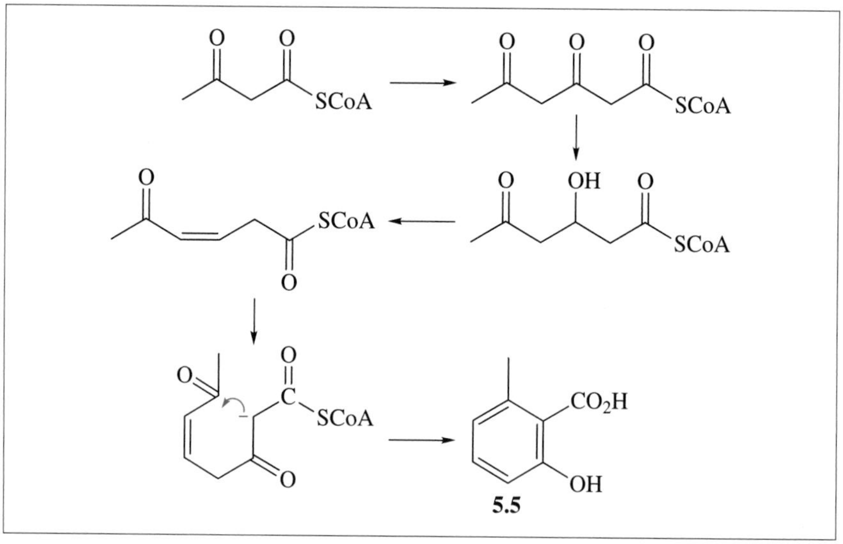

Scheme 5.4   The formation of 6-methylsalicylic acid

The use of carbon-13 NMR spectroscopy has been of major importance in defining the folding of polyketide chains. An example of these methods comes from work on the metabolites of the fungus *Aspergillus melleus*. This fungus produces a group of pentaketide metabolites including mellein (**5.6**) and aspyrone (**5.7**). When mellein was biosynthesized from $[1,2\text{-}^{13}C_2]$acetate, it showed the expected five pairs of $^{13}C\text{-}^{13}C$ couplings (shown as thick bonds). However, the branched chain of aspyrone was more difficult to rationalize. Carbons 1, 3, 5 and 8 were enriched by $[1\text{-}^{13}C]$acetate and carbons 2, 4, 6, 7 and 9 were enriched by $[2\text{-}^{13}C]$acetate. When $[1,2\text{-}^{13}C_2]$acetate was the substrate, there were three pairs of couplings involving C(2)–C(3), C(4)–C(5) and C(8)–C(9). In addition, there was a small two-bond coupling between C(1) and C(7), indicating that these two carbons had their origin in the same acetate unit which had undergone an intramolecular rearrangement (see **5.8**) in the course of the biosynthesis.

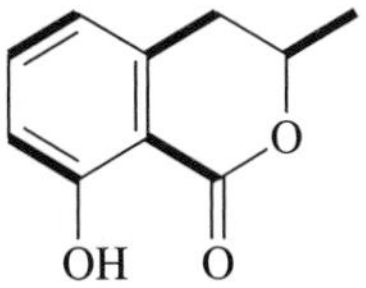

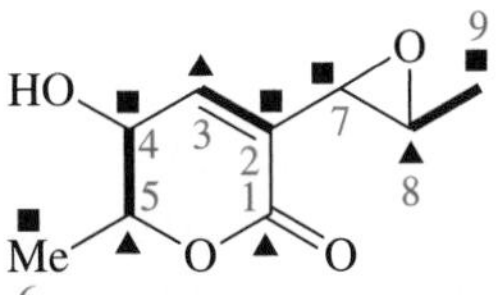

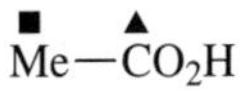

The further modification of polyketides by the insertion of methyl groups from methionine, of isoprene units from isopentenyl pyrophosphate, and by the cleavage of the aromatic ring and by oxidative coupling affords a diverse range of natural products. The oxidative phenol coupling of griseophenone B (**5.9**) to 4-demethyldehydrogriseofulvin (**5.11**), and the transformation of the latter to griseofulvin (**5.10**), is an example (see Scheme 5.5).

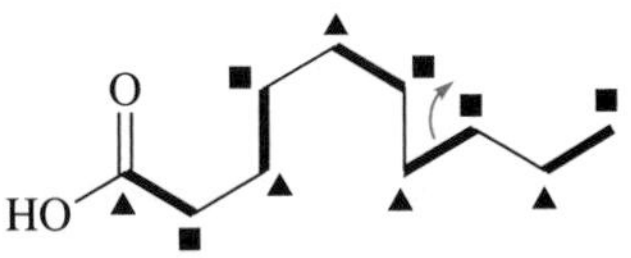

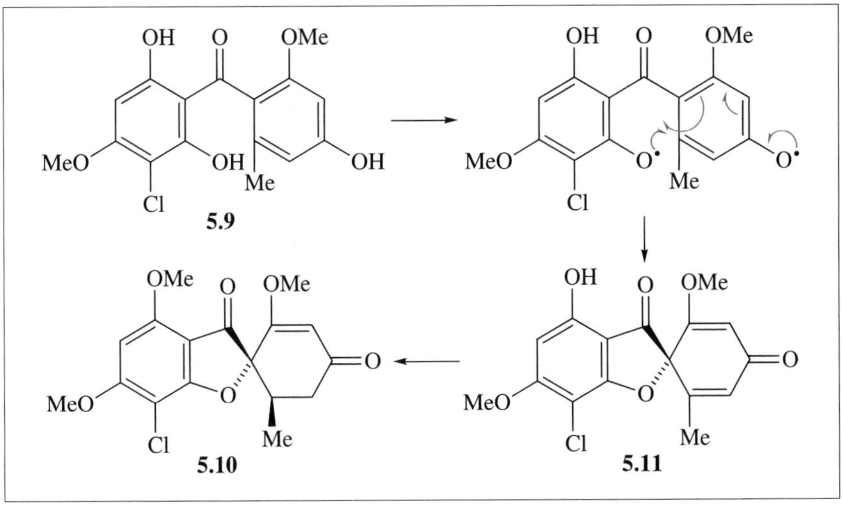

Scheme 5.5 The biosynthesis of griseofulvin

5.5   Terpenoid Biosynthesis

Although the "isoprene rule" as an aid for structure elucidation was established in the 1920s by Ruzicka, the biosynthetic origin of the C_5 unit was unknown. There are now known to be two major pathways, one based on **mevalonic acid** (**5.13**; Scheme 5.6) and the other on **1-deoxyxylulose** (**5.16**; Scheme 5.7).

Studies on the biosynthesis of **cholesterol** (**5.24**; Scheme 5.11) from acetate showed that the isoprene unit in mammals was constructed from three acetate units with the loss of one carboxyl carbon.

A breakthrough came in 1956 with the demonstration that mevalonic acid (**5.13**), which had been identified as growth-promoting factor for a bacterium, *Lactobacillus acidophilus*, was efficiently incorporated into cholesterol by rat liver slices. It was subsequently shown to be specifically incorporated into many other terpenoid substances, particularly by fungi.

Terpenoid biosynthesis can be divided into four stages. Firstly, there is the formation of the isoprene unit **isopentenyl pyrophosphate** (**5.14**; Scheme 5.6); secondly, there is the association of these units to form the $(C_5)_n$ isoprenoid backbone of the terpenoid families; thirdly, there is the cyclization of these to generate the carbon skeletons; and finally, there are the interrelationships, hydroxylations and oxidations that lead to the individual terpenoids.

NADH is a co-enzyme. It is the reduced form of nicotinamide adenine dinucleotide and acts as a biological reducing agent.

The first irreversible step in terpenoid biosynthesis on the mevalonate pathway (Scheme 5.6) involves the enzymatic reduction of (S)-3-hydroxy-3-methylglutaryl co-enzyme A (**5.12**, HMG-CoA) with hydrogen from nicotinamide adenine dinucleotide to produce (R)-mevalonic acid (**5.13**). The HMG-CoA arises mainly by the condensation of acetyl co-enzyme A with acetoacetyl co-enzyme CoA. Two successive phosphorylations of mevalonic acid produce the 5-pyrophosphate.

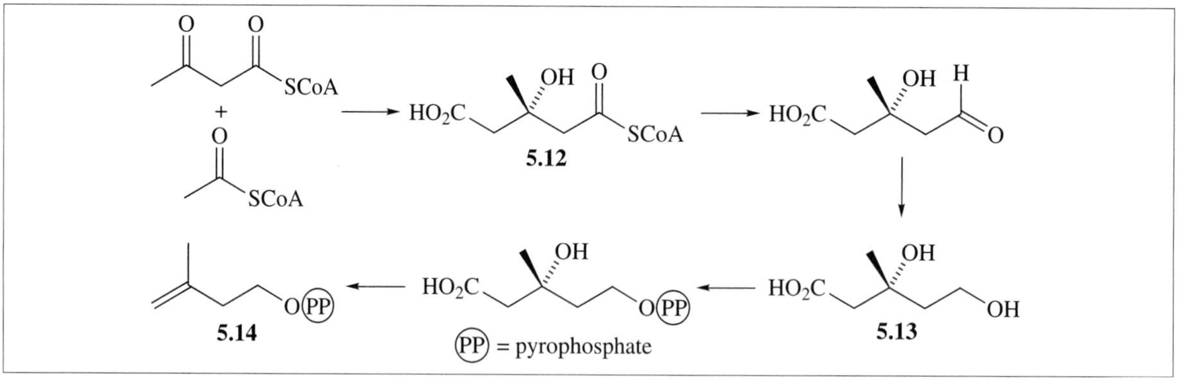

Scheme 5.6   The mevalonate pathway to isopentenyl pyrophosphate

This undergoes a *trans* elimination of the tertiary hydroxyl group and the carboxyl group to form 3-methylbut-3-enyl pyrophosphate (isopentenyl pyrophosphate, IPP, **5.14**). Careful labelling studies by Cornforth established the stereochemistry of this and many other steps in terpenoid biosynthesis.

For many years it was difficult to obtain definitive evidence for the intervention of mevalonic acid in the biosynthesis of terpenoids in some *Streptomycetes* and in some plants. Recently, a pathway based on the formation of 1-deoxyxylulose (**5.16**) has come to light. The full details of this pathway have yet to be established, but it is shown in outline in Scheme 5.7. Thiamine pyrophosphate (**5.15**) plays an important role in the condensation of pyruvic acid (2-oxopropanoic acid) and glyceraldehyde monophosphate.

Thiamine (vitamin B_1) pyrophosphate is a co-enzyme which can mediate the transfer of C_2 units in biosynthesis.

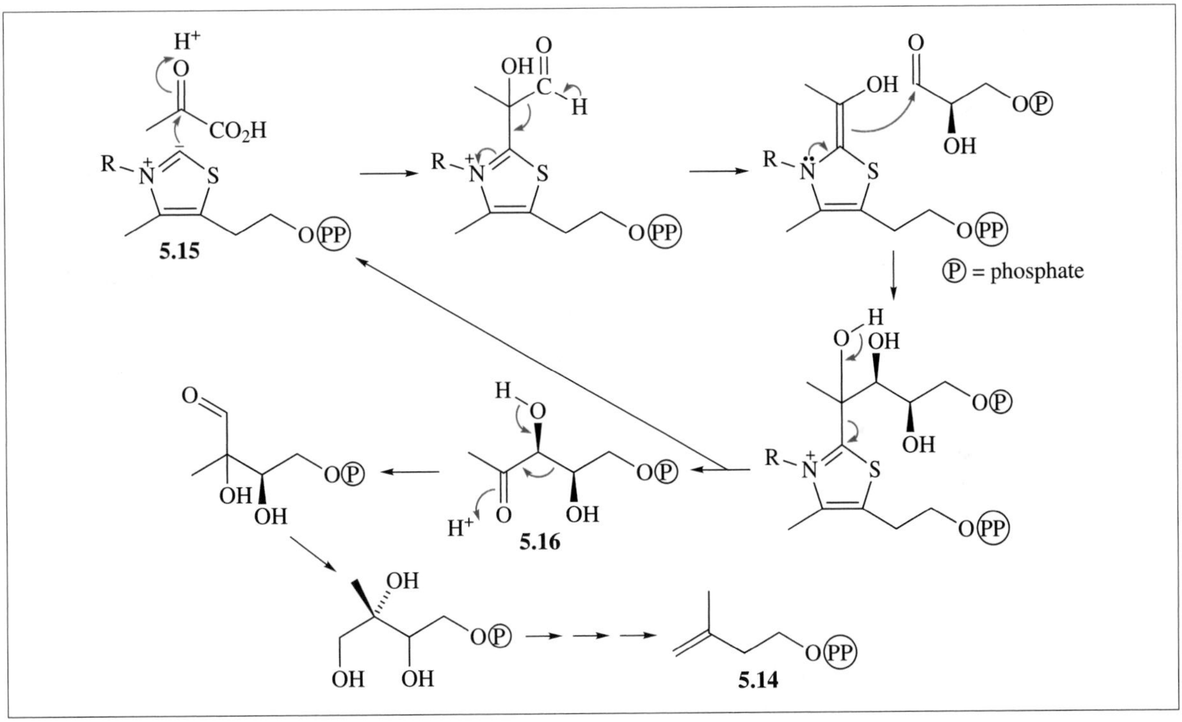

Scheme 5.7 The 1-deoxyxylulose pathway to isopentenyl pyrophosphate

A stereospecific but reversible isomerization of the double bond of IPP (**5.14**) produces **3-methylbut-2-enyl pyrophosphate** (dimethylallyl pyrophosphate, DMAPP, **5.18**) (Scheme 5.8). The significance of this isomerization is the creation of a reactive allylic pyrophosphate which plays an important role in the association of two isoprene units to form the C_{10} **geranyl pyrophosphate** (**5.17**). The allylic pyrophosphate is used in an enzyme-catalysed (prenyl transferase) alkylation reaction of IPP. The loss

In a reactive allylic pyrophosphate, the double bond can provide resonance stabilization of an adjacent carbocation.

of a pyrophosphate anion to form the alkylating unit together with a proton from the IPP generates a further reactive allylic pyrophosphate. Labelling experiments have shown that the formation of the new carbon–carbon bond between the isoprene units is accompanied by inversion of configuration at the allylic carbon atom. The stereochemistry of the loss of the proton from the IPP suggests that this alkylation reaction may take place in two steps. It is important to note that when an intermediate is bound to a chiral enzyme surface, two apparently identical hydrogens lose their identity and may be distinguished by the enzyme. Chirally labelled mevalonates have played a valuable role in establishing the stereochemistry of these reactions. The addition of further isoprene units gives farnesyl pyrophosphate (**5.19**, C_{15}) and geranylgeranyl pyrophosphate (C_{20}), which are the parents of the sesqui- and diterpenoids.

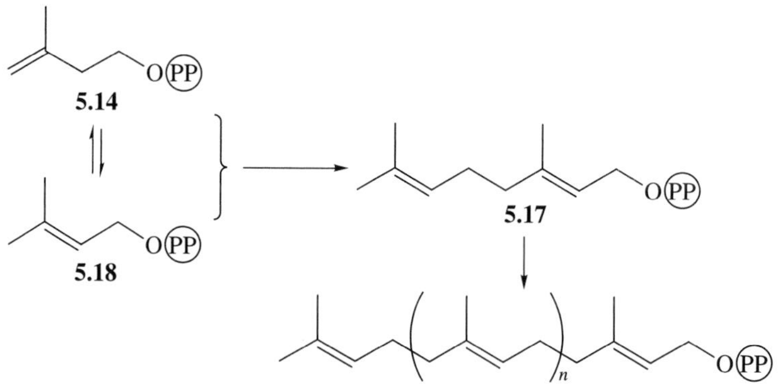

5.19 $n = 1$ Farnesyl pyrophosphate (C_{15})
$n = 2$ Geranylgeranyl pyrophosphate (C_{20})

Scheme 5.8 The formation of polyprenyl chains

Two molecules of **farnesyl pyrophosphate** are joined together in a "tail-to-tail" manner to form **squalene** (**5.21**, C_{30}). This is the parent of the triterpenes and steroids. Phytoene (C_{40}), the parent of the carotenoids, arises from two geranylgeranyl pyrophosphate units. In the course of the coupling of the two farnesyl pyrophosphate units, a hydrogen atom from one unit is stereospecifically replaced by a hydrogen originating from NADPH. The mechanism of the coupling (Scheme 5.9) involves a cyclopropyl intermediate, presqualene pyrophosphate (**5.20**). This arises by an alkylation of the double bond of one farnesyl unit by the pyrophosphate of the other. Rearrangement and reduction leads to squalene (**5.21**).

NADPH is the reduced form of nicotinamide adenine dinucleotide phosphate.

The cyclization of **squalene epoxide** (**5.22**) to form the triterpene **lanosterol** (**5.23**) involves a series of stereospecific cyclizations and

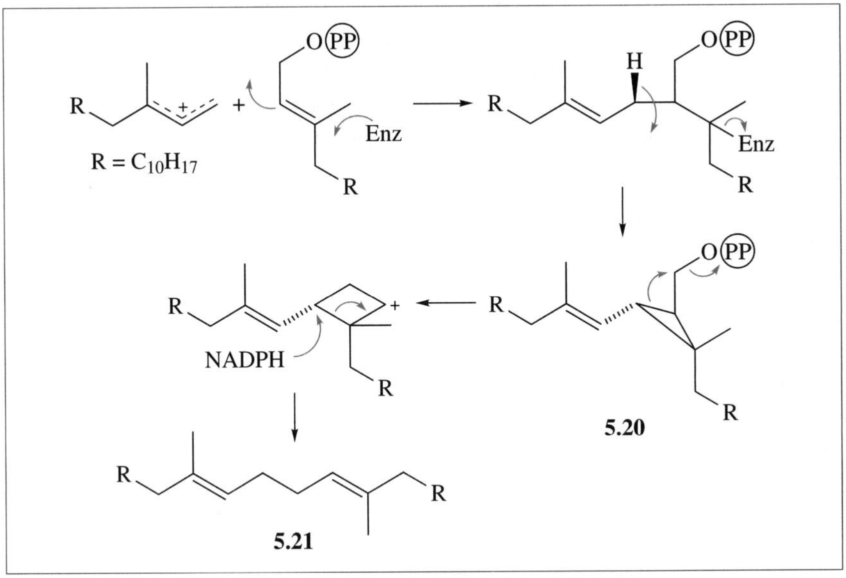

Scheme 5.9 The formation of squalene

rearrangements. At one time it was thought that the whole process took place in one reaction. However, recent evidence suggests that the cyclization is not fully concerted and there may be some discrete intermediates, as shown in Scheme 5.10. A five-membered ring C is formed by a Markownikoff ring closure. This then undergoes a ring expansion before the final stages of the cyclization and rearrangement take place.

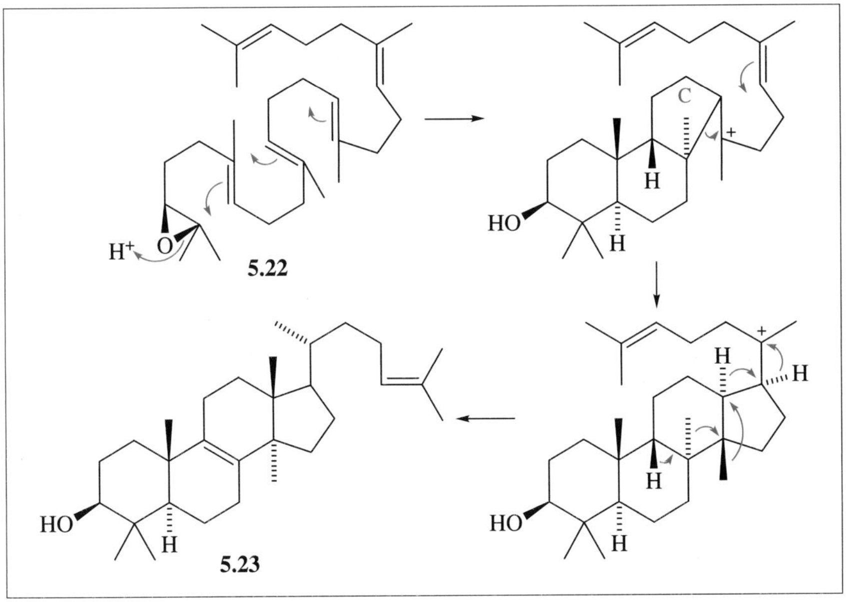

Scheme 5.10 The cyclization of squalene epoxide

The conversion of lanosterol (**5.23**) to **cholesterol** (**5.24**), and subsequently to other sterols and steroid horomones, follows a well-defined sequence of events. Some steps in this pathway are set out in Scheme 5.11. The demethylation stages involve the oxidation of the C-4 methyl groups to a carboxylic acid and the decarboxylation of a β-keto acid. However, the loss of the C-14 methyl group takes place after it has been oxidized to an aldehyde. This particular step is the target for inhibition by the azole fungicides, in which the nitrogen heterocycles coordinate to iron atoms responsible for the oxidation. The modification and removal of the side chain of cholesterol leads to the different families of steroid. One late stage which has attracted considerable interest is the oxidative removal of C-19 and the conversion of ring A of the steroids to a phenol, as in **estradiol** (**5.25**). A significant proportion of breast cancers are estrogen dependent. Hence this enzyme system, known as aromatase, has been a target for cancer chemotherapy.

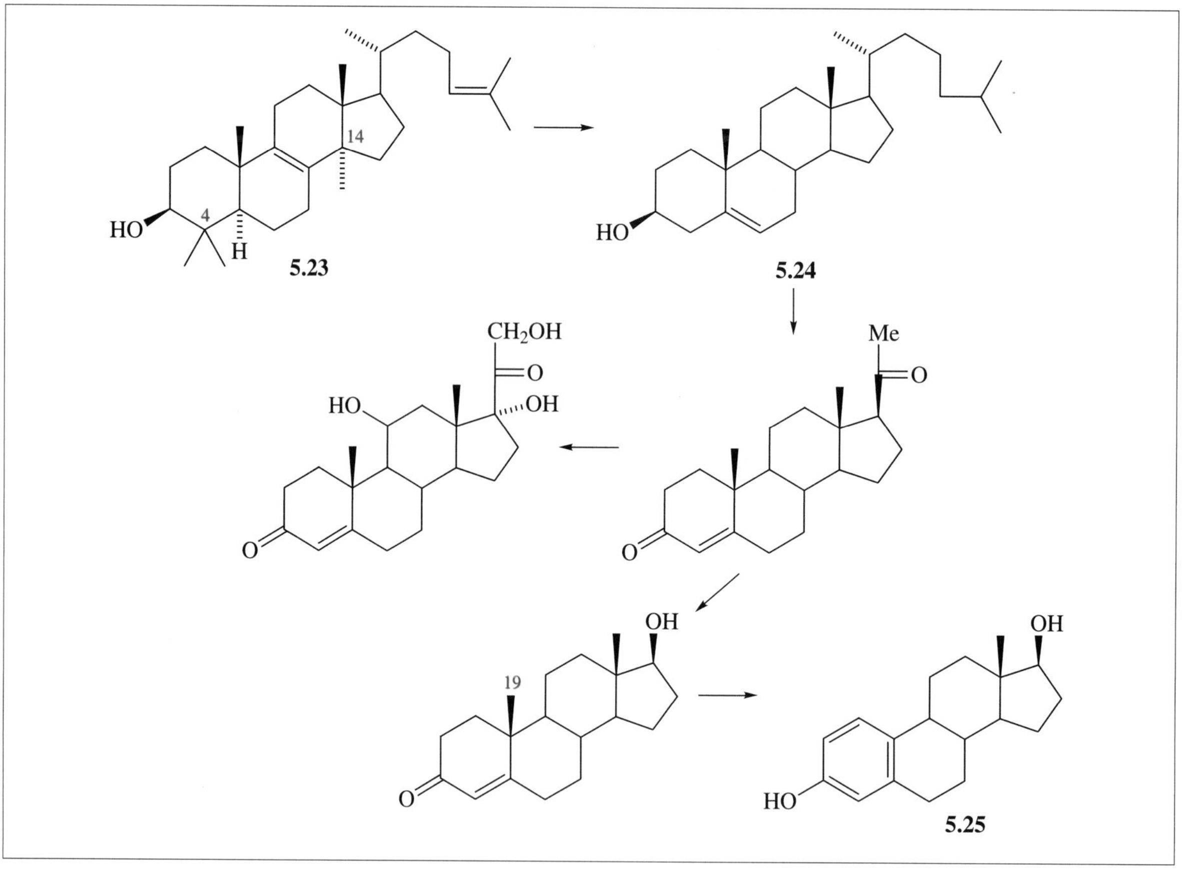

Scheme 5.11   The formation of the steroid hormones

There are two general modes of cyclization of the acyclic terpenoid precursors. In the first, the cyclization is initiated by protonation of an alkene or, as in the case of the triterpenoids, an epoxide. In the second mode, the cyclization is initiated by electrophilic attack of the allylic pyrophosphate or the isomeric pyrophosphate of a 3-hydroxy-1-ene on a double bond in another isoprene unit. In the monoterpenoids, this can generate the menthene skeleton (Scheme 5.12), and in the sesquiterpenoids it can lead to the medium-sized rings of the humulene series. In the diterpenoids it leads to the formation of the macrocyclic ring of the cembrenes. Further cyclizations of these then lead to a wide range of different skeletons. The isolation of the enzyme systems responsible for many of these terpene cyclizations is of considerable value in the study of their mechanism.

Humulene is a sesquiterpene that is found in hops. Many other sesquiterpenes are formed by the further cyclization of humulene carbocations.

Cembrene is a diterpene that is found in pines and in tobacco.

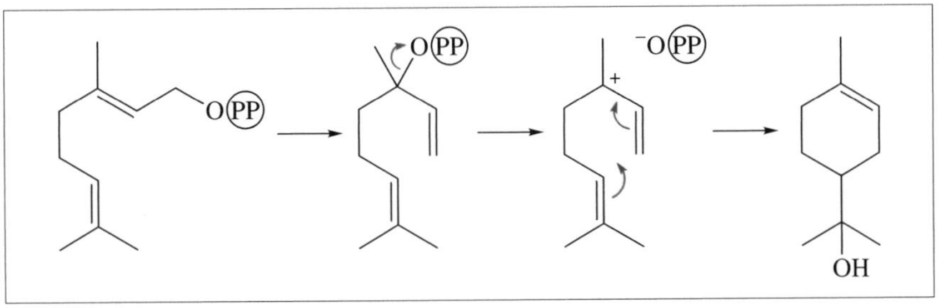

Scheme 5.12 The cyclization of geranyl pyrophosphate

Some of the sequences of hydroxylations and oxidations that lead to the individual terpenoids, such as the gibberellin plant growth hormones, have been examined in detail (see Scheme 5.13). The **gibberellins** have a very powerful influence on plant growth and development, and as a result their biosynthesis has been thoroughly studied. They are present in plants in very low concentrations ($\mu g/g$). However, the fungus *Gibberella fujikuroi* produces some gibberellins in large amounts, and it formed a useful organism for studying their biosynthesis. A number of plant enzyme systems have now been cloned and expressed in bacteria such as *Escherichia coli* so that they are readily available. The gibberellins are formed by the stepwise oxidation of a tetracyclic diterpenoid hydrocarbon, *ent*-**kaurene** (**5.28**). This hydrocarbon is formed by a two-stage cyclization of the C_{20} geranylgeranyl pyrophosphate (**5.26**). Oxidation of the hydrocarbon at C-19 and C-7 gives *ent*-7α-hydroxykaurenoic acid (**5.27**), which is the substrate for a unique oxidative ring contraction to form **gibberellin A$_{12}$ 7-aldehyde** (**5.29**). Several of these earlier steps are targets for plant growth regulators. There are then parallel pathways which take place in different species involving hydroxylation at various centres, *e.g.* C-3, and which lead to particular families of gibberellins. However, a key stage in all these pathways is the loss of C-20 and the formation of the γ-lactone typical of the biologically active hormones such as **gibberellic acid** (**5.30**).

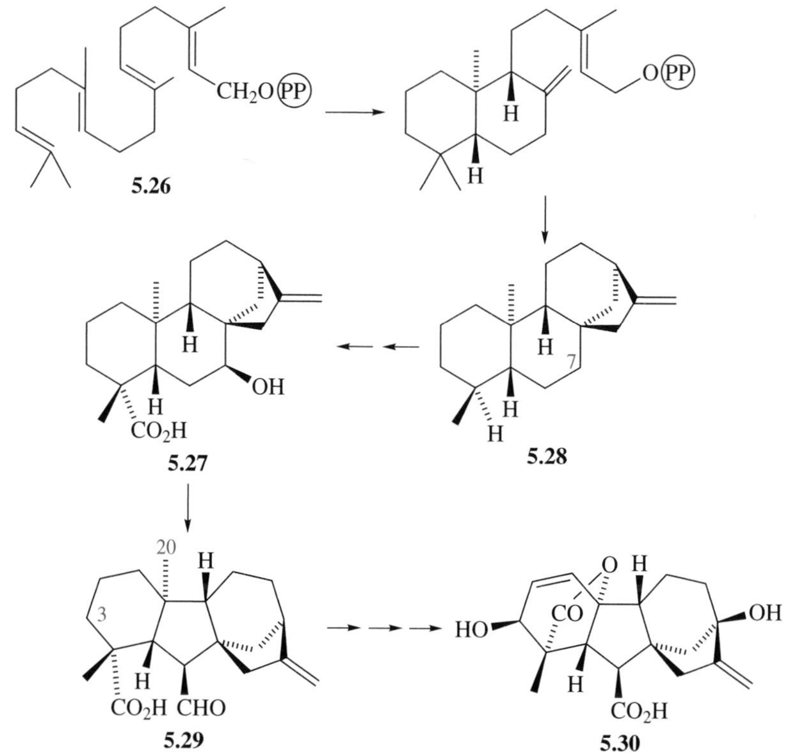

Scheme 5.13 The formation of the gibberellin plant hormones

Worked Problem 5.1

Q   The labelling pattern of the sesquiterpenoid illudin M (**1**), obtained from the fungus *Clitocybe illudens*, where it is biosynthesized from [1-^{13}C]acetate and [1,2-^{13}C$_2$]acetate, is as shown. Show how farnesyl pyrophosphate may be folded to generate this labelling pattern.

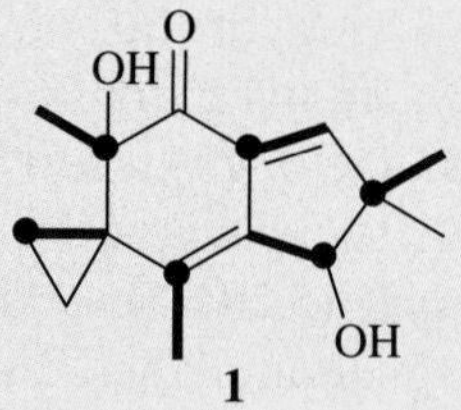

1

— denotes pairs of coupled atoms

● denotes atoms enriched by [1-^{13}C]acetate

A   An isoprene unit is derived via mevalonic acid from three molecules of acetate and it is therefore labelled by acetate as in **2**. Three isoprene units are incorporated into farnesyl pyrophosphate (**3**). A folding of farnesyl pyrophosphate which generates the labelling pattern of illudin M is shown in **4**.

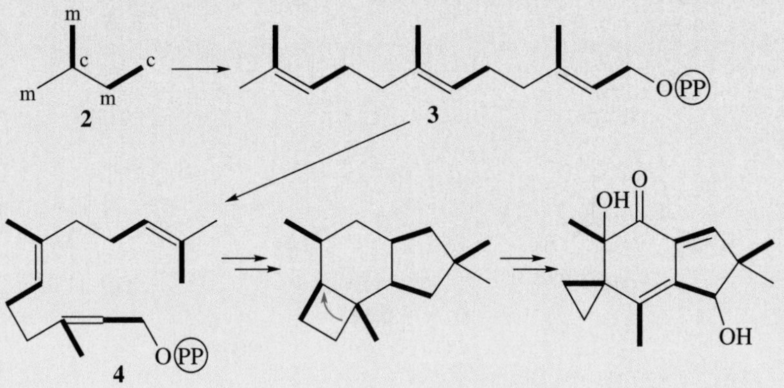

5.6   The Biosynthesis of Phenylpropanoids

The existence of a large group of secondary metabolites containing an aromatic ring and a three-carbon side chain, the phenylpropanoid C_6–C_3 series, was recognized many years ago. An important key to their biosynthesis came from an examination of mutants of the bacterium *Escherichia coli* which would grow only on a medium supplemented with the C_6–C_3 amino acids phenylalanine or tyrosine. It was found that **shikimic acid** (**5.34**) was an alternative non-aromatic supplement that facilitated the growth of these mutants. The study of this pathway fell into two sections, the formation of shikimic acid and the conversion of this to the aromatic C_6–C_3 compounds. Several of the steps in the pathway may be rationalized in terms of enolate chemistry. The sequence which has been established by labelling studies is set out in Scheme 5.14.

The initial step involves the enzyme-catalysed condensation of a derivative of phosphoenol pyruvate (**5.31**) with the aldehyde of erythrose 4-phosphate (**5.32**) to form a C_7 acid, **heptulosonic acid** (**5.33**). Elimination of the phosphate and a further enolate condensation gives **3-dehydroquinic acid** (**5.35**). The overall stereochemistry of the elimination of water to give **dehydroshikimic acid** has been shown to involve a "*syn*" process and this may therefore be a stepwise sequence involving an enolization and elimination. Reduction of the 3-dehydroshikimic acid yields **shikimic acid** (**5.34**).

Although the aromatic amino acids might be considered to be primary metabolites, this pathway plays a major role in the formation of many secondary metabolites.

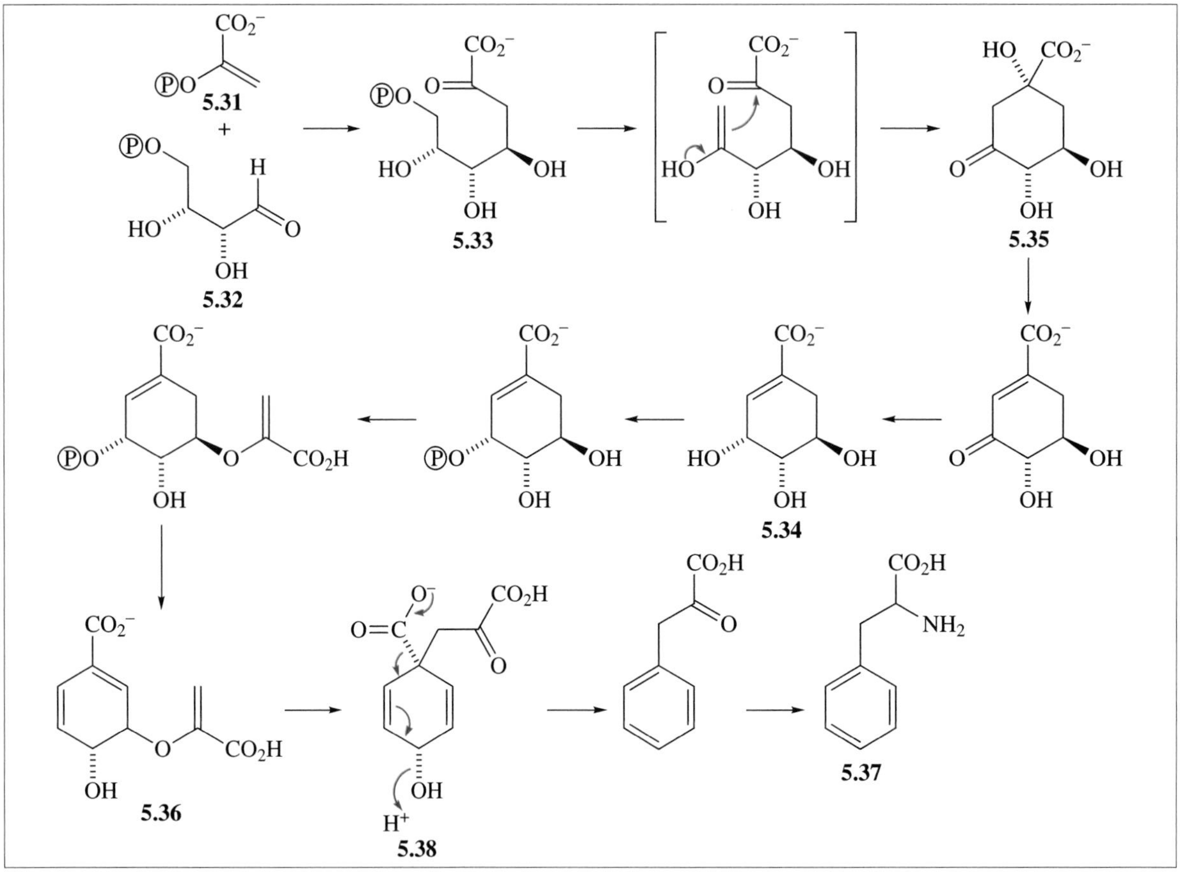

Scheme 5.14 The biosynthesis of C$_6$–C$_3$ compounds

Amino nitrogen is introduced via pyridoxamine in a process known as transamination. This involves a carbonyl–amine condensation. Phenylalanine is formed from phenylpyruvic acid.

The addition of the three-carbon side chain to shikimic acid 3-phosphate to give **chorismic acid** (**5.36**) starts with the formation of an enol ether from phosphoenol pyruvate. The stereochemistry of the elimination of phosphoric acid to form chorismic acid has been shown to be an *anti* 1,4-elimination and it is possible that this also occurs in a stepwise manner. The stereochemistry of the Claisen rearrangement of chorismic acid (**5.36**) to **prephenic acid** (**5.38**) by the enzyme chorismate mutase has been thoroughly studied. The conversion of prephenic acid to **phenylalanine** (**5.37**) or to tyrosine involves both a decarboxylation and a transamination. Since this pathway occurs in plants but not in man, it has been the subject of a number of studies in the design of herbicides.

There are a number of families of natural products that are derived from the shikimic acid pathway. The oxygenation pattern of an aromatic ring derived from this pathway differs from that of polyketide aromatic rings. The C$_6$–C$_3$ 4-hydroxycinnamoyl unit can act as a "starter" for a polyketide biosynthesis, as in the formation of the flavanoid plant pigments. The oxygenation patterns of the rings of these compounds (*e.g.* **5.39**) reveals their different biosyntheses.

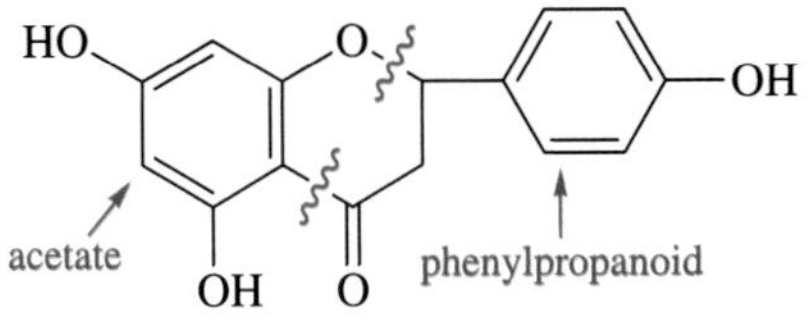

5.39

Chorismic acid (**5.36**) is also the source of anthranilic acid (2-aminobenzoic acid, **5.41**) and the amino acid tryptophan (**5.42**) (Scheme 5.15). Anthranilate synthase mediates an addition of ammonia and an elimination reaction. Condensation of the anthranilic acid (**5.41**) with ribose 5-phosphate, followed by an isomerization and decarboxylation, leads to indole-3-glycerol phosphate (**5.40**) and thence the amino acid tryptophan (**5.42**).

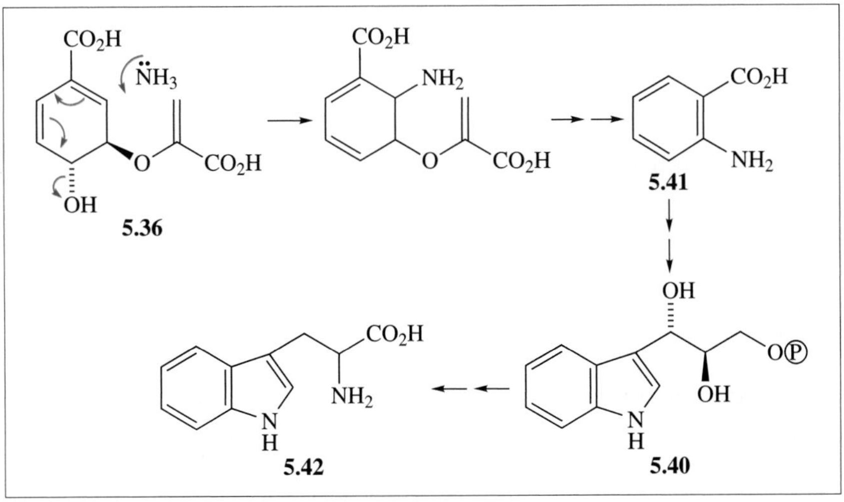

Scheme 5.15 The formation of tryptophan

5.7   Alkaloid Biosynthesis

Alkaloids are a structurally diverse family of nitrogen-containing bases. A unifying feature is the biosynthetic origin of the majority from a limited number of amino acids. Alkaloids may be classified into those that derived from: (i) lysine or ornithine, (ii) phenylalanine or tyrosine and (iii) tryptophan. In addition, there are those alkaloids, such as the steroidal alkaloids, in which the nitrogen is introduced after the final carbon skeleton has been formed.

The basic amino acids ornithine (**5.41**) and lysine (**5.42**) are the precursors of the pyrrolidine and piperidine rings that are found in a number of alkaloids. An important aspect of these biosyntheses is the role of the iminium ion in a Mannich type of condensation. The alkaloid

Iminium ions may react with enols or enamines in C–C bond formation.

nicotine (**5.43**) is biosynthesized from nicotinic acid by a reaction of this type (see Scheme 5.16).

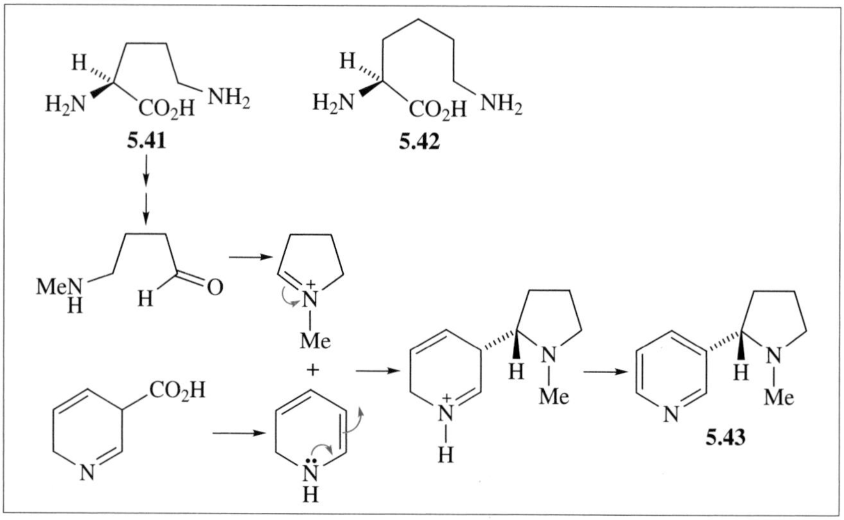

Scheme 5.16   The biosynthesis of nicotine

A large number of alkaloids are derived from the amino acid **tyrosine** (**5.44**). Amongst these the biosynthesis of **morphine** has attracted considerable interest. The alkaloids that are found in the opium poppy fall into two groups: the benzylisoquinoline series such as papaverine and laudanosoline, and the more complex alkaloids such as codeine and morphine.

Three pieces of **biogenetic speculation** guided the biosynthetic studies on morphine. The first was the suggestion, by Wintersteiner and Trier and by Robinson, of the role of amino acids in the biosynthesis of the benzylisoquinoline alkaloids. Secondly, the co-occurrence of the benzyl-isoquinoline and morphine alkaloids led to the suggestion by Robinson of a structural and biogenetic relationship. Rotation of the benzylisoquino-line alkaloids about the axis shown in Scheme 5.17 reveals a possible relationship to the morphine skeleton. Thirdly, the suggestion was made by Barton that phenol coupling played an important role in the formation of the morphine ring system.

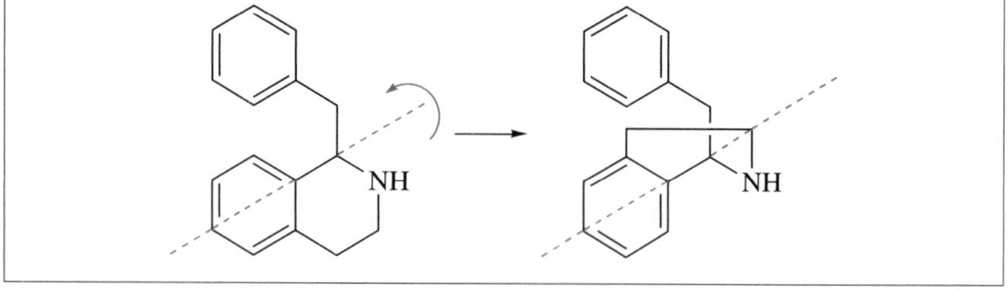

Scheme 5.17   The relationship of the benzylisoquinoline and morphine alkaloids

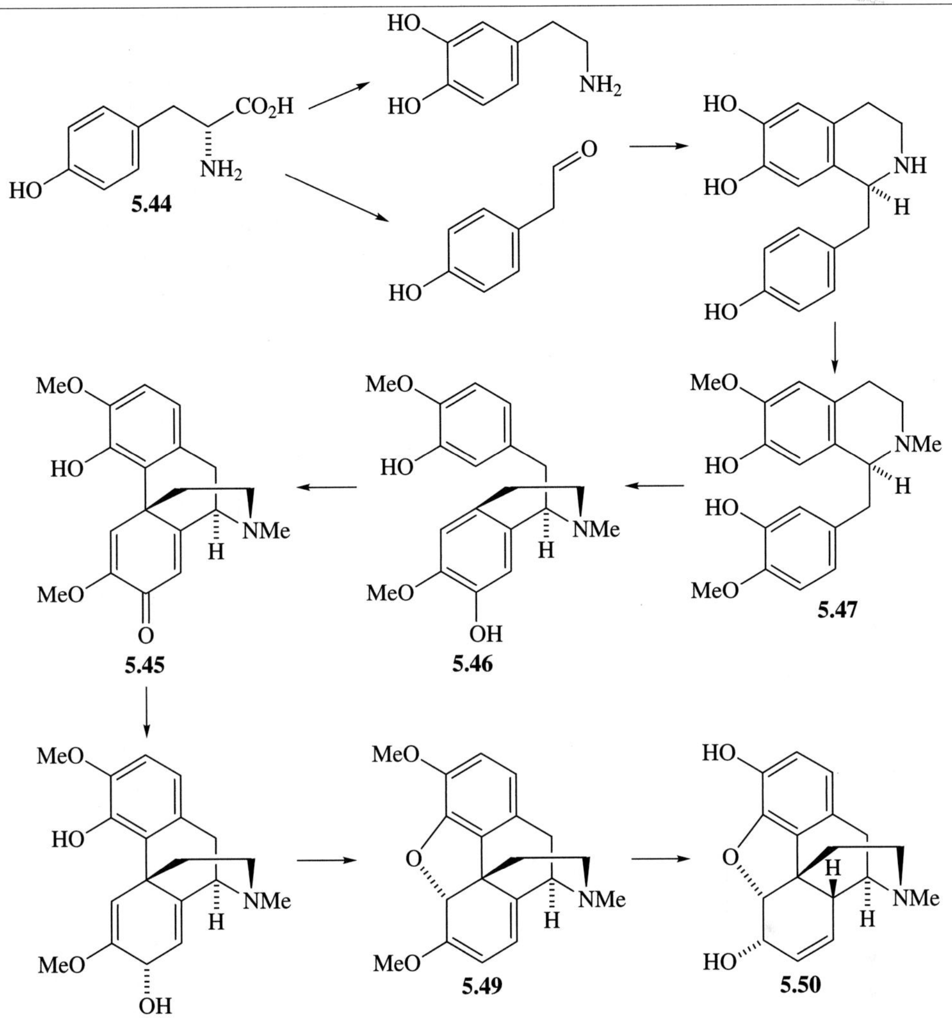

Scheme 5.18 The biosynthesis of morphine

Biosynthetic experiments by Barton, Battersby and others have established the pathway outlined in Scheme 5.18. There are several interesting steps. Firstly, both the benzyl and the isoquinoline rings originate from tyrosine (**5.44**). Secondly, the (*S*)-enantiomer of the benzylisoquinoline **reticuline** (**5.47**) is formed first and this is isomerized to (*R*)-reticuline (**5.46**) prior to the phenol-coupling step and the conversion to **salutaridinone** (**5.45**). Multiple-labelling experiments established the significance of the methylation pattern of reticuline, which directs the phenol coupling. The formation of the ether bridge of morphine requires an elimination reaction involving **salutaridinol** (**5.48**). The sequence **thebaine** (**5.49**), **codeine** and then **morphine** (**5.50**) was established by briefly exposing the plants to an atmosphere of radioactive carbon dioxide and following the appearance of radioactivity in the alkaloids.

Phenol coupling plays a role in generating the carbon skeleton of a number of other alkaloids derived from tyrosine, such as galanthamine which is found in daffodils.

The large family of indole alkaloids include reserpine and strychnine and the dimeric alkaloids such as vincaleucoblastine. These alkaloids have been shown to be derived from the amino acid tryptophan and a C_{10} unit. The isoprenoid origin of the latter from the monoterpenoid *seco*-loganin was established by labelling studies.

Worked Problem 5.2

Q   Indicate the sites that are labelled in morphine when it is biosynthesized from $[2\text{-}^{14}C]$tyrosine.

A   The biosynthesis of morphine is set out in Scheme 5.18. Two molecules of tyrosine are involved in its biosynthesis *via* reticuline. This leads to the labelling pattern as shown below:

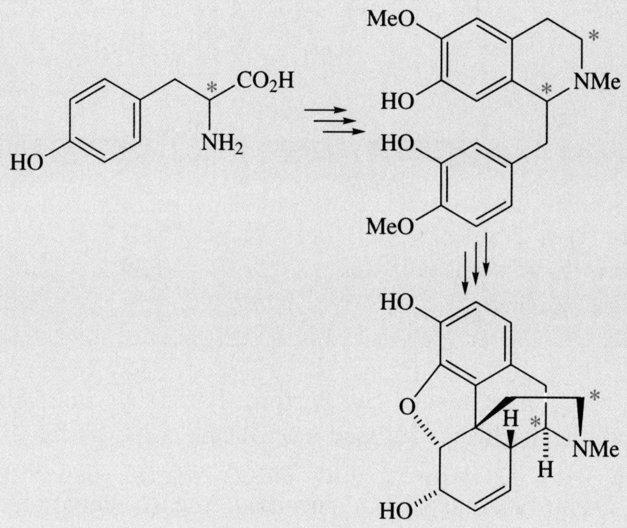

5.8   Other Natural Products Derived from Amino Acids

A number of other important families of natural products are derived from amino acids. The **penicillins** are an example (Scheme 5.19). The penicillins

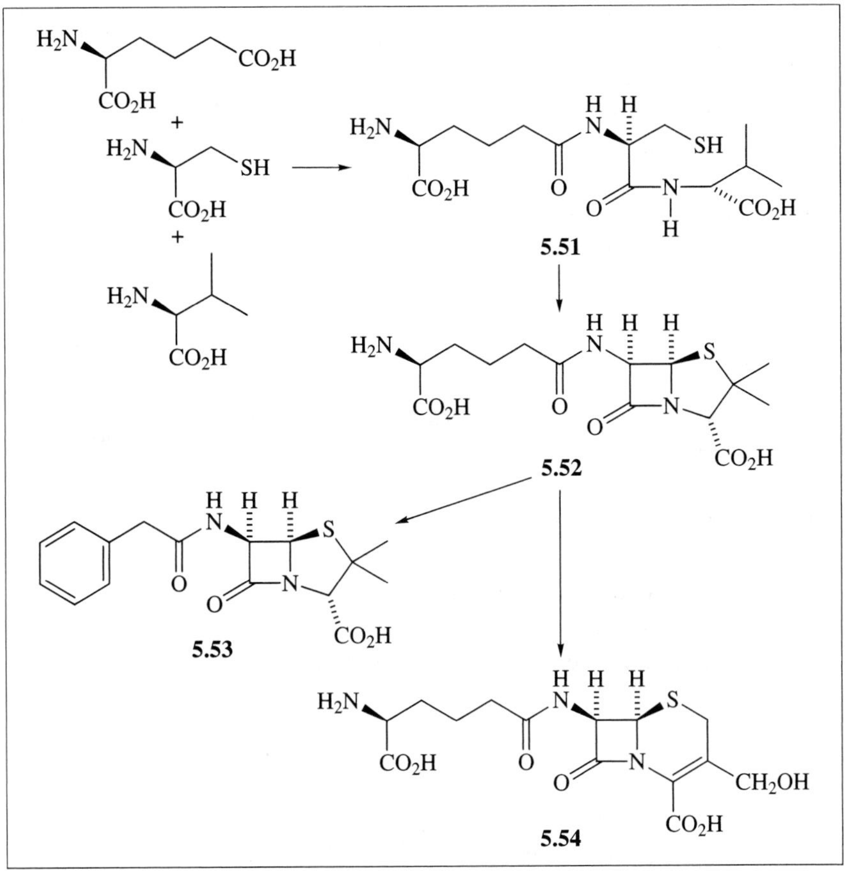

Scheme 5.19 The biosynthesis of penicillins and cephalosporins

are derived from a tripeptide, δ-(L-α-aminoadipoyl)-L-cysteinyl-D-valine (**5.51**), in which the valine unit possesses the unusual D configuration. This tripeptide is cyclized oxidatively by an enzyme system known as isopenicillin N synthase to form **5.52**. The mechanism of this cyclization has attracted considerable interest. The isopenicillin N is converted to the penicillins, such as **5.53**, and to pencillin N which in turn undergoes a ring expansion to form the cephalosporins (**5.54**).

Chlorophyll, which is essential for photosynthesis, haem, the oxygen-containing component of haemoglobin, the cytochromes and vitamin B_{12} all contain a tetrapyrrole nucleus with various attachments and different central metals (magnesium, iron or cobalt). However, the biosynthesis of these so called "pigments of life" share some common features in that they have a single biosynthetic parent known as uroporphyrinogen III, (uro'gen III, **5.57**). Once this has been formed (Scheme 5.20), it is subjected to considerable biosynthetic modification to give the individual substances. Uro'gen III contains four pyrrole rings and is biosynthesized from a pyrrole, porphobilinogen (PBG, **5.56**). This arises from

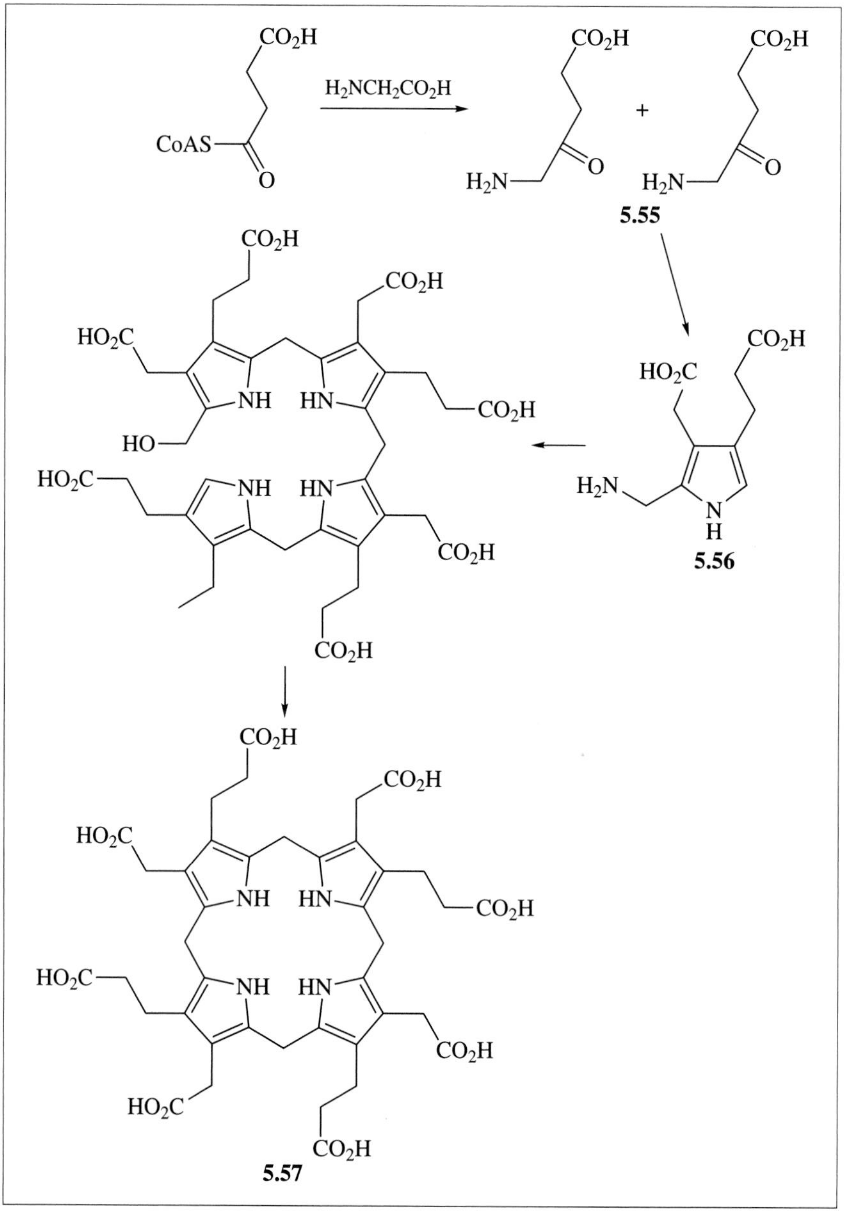

Scheme 5.20 The formation of uroporphyrinogen III

the condensation of two molecules of the amino acid 5-aminolevulinic acid (ALA, **5.55**). The simple linear head-to-tail combination of four porphobilinogen units would lead to a tetrapyrrole, called a bilane, with alternating acetate and propionate side chains. However, this is not the case in uro'gen III. The four molecules of porphobilinogen are first brought together in a linear manner to give a tetrapyrrole by an enzyme system known as **deaminase** (the amine is lost in this process). A second enzyme system, **cosynthetase**, carries out the ring closure of the bilane with rearrangement to form uro'gen III. The detailed stereochemistry and mechanism of the various steps, such as the *C*-methylations and ring contraction that lead to vitamin B_{12} and the changes that generate the other pigments, are outside the scope of this book. However, the underlying theme is that of many biosyntheses in which a common building block is assembled in a general manner and then specifically modified to give the individual natural products.

Summary of Key Points

1. Labelling studies using radioisotopes (^3H and ^{14}C) and stable isotopes (^2H, ^{13}C and ^{18}O) together with enzymatic methods have been used to establish biosynthetic pathways.

2. Polyketides, including fatty acids and some aromatic compounds, are derived from the linear condensation of acetate (malonate) units. Fatty acid biosynthesis involves a sequence of condensations and reductions. Aromatic compounds are formed by aldol-type condensations. The biosynthesis takes place in a multi-enzyme complex known as polyketide synthase (PKS).

3. The prostaglandins are formed from arachidonic acid by cyclo-oxygenase.

4. The C_5 building block of the terpenoids and steroids is isopentenyl pyrophosphate. This may be formed from mevalonic acid or from 1-deoxyxylulose.

5. The steroids are formed by the cyclization of squalene epoxide to lanosterol and the conversion of the latter to cholesterol.

6. Phenylpropanoid secondary metabolites are biosynthesized via shikimic acid. The aromatic ring has a different oxygenation pattern from those compounds that are biosynthesized from acetate units.

7. Alkaloids are biosynthesized from amino acids such as tyrosine. The biosynthesis of morphine includes a phenol coupling reaction involving a benzylisoquinoline alkaloid, reticuline.

8. Penicillin is biosynthesized by the oxidative cyclization of a tripeptide, L-aminoadipoyl-L-cysteinyl-D-valine. The tetrapyrrole ring system, as in chlorophyll, heme and vitamin B_{12}, is formed from porphobilinogen.

Problems

5.1. Mark with an asterisk the sites that might be labelled in the following natural products when they are biosynthesized from the labelled precursors that are given:

(a) 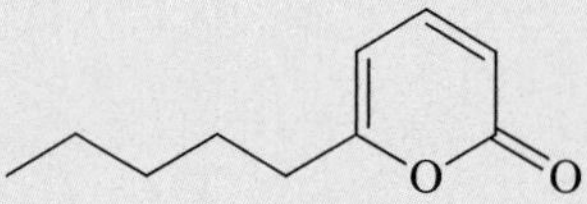   from   Me*CO$_2$H

(b)   from   HO$_2$C

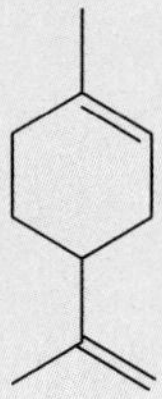

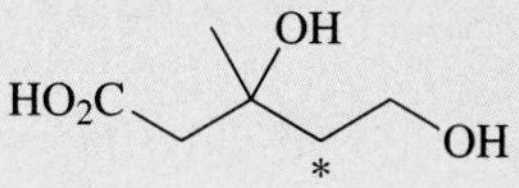

(c) 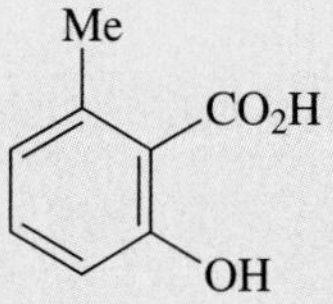   from   Me*CO$_2$H

(d) 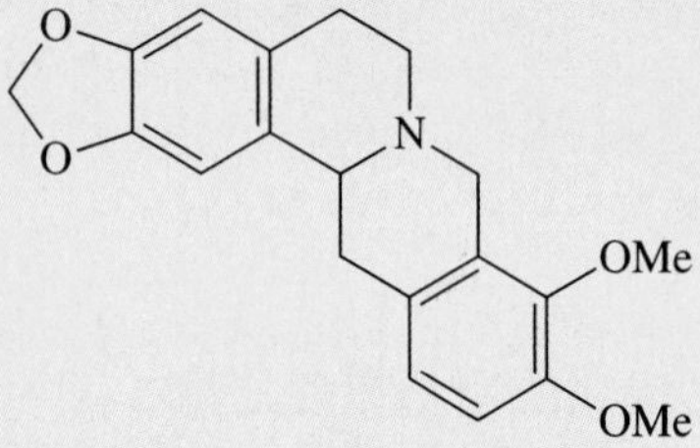   from

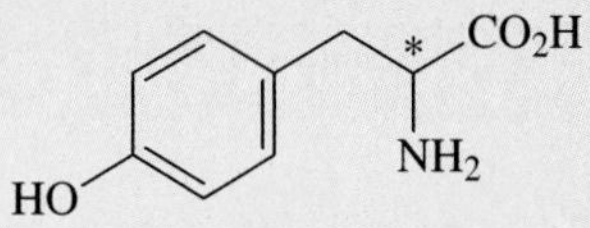

(e) 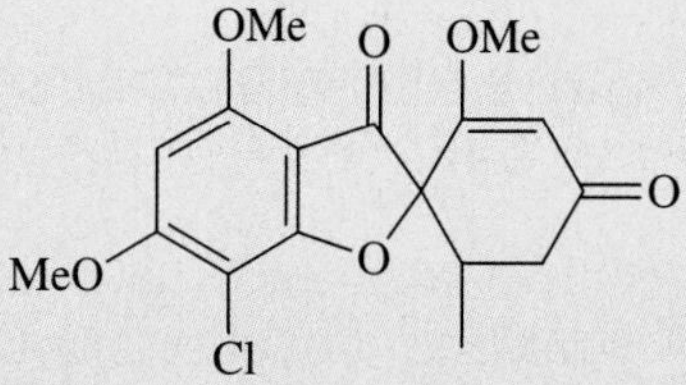   from   Me*CO$_2$H

5.2. Indicate the biosynthetic structural units [acetate, C$_1$ (methionine), C$_5$ (isoprene), C$_6$–C$_3$ (shikimate)] that form the carbon skeletons of the following compounds. Some of these compounds may be formed by the combination of more than one pathway.

(a)

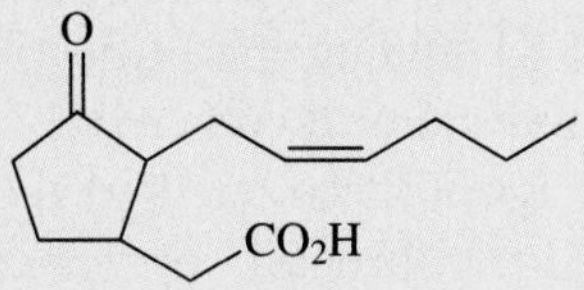

(b)

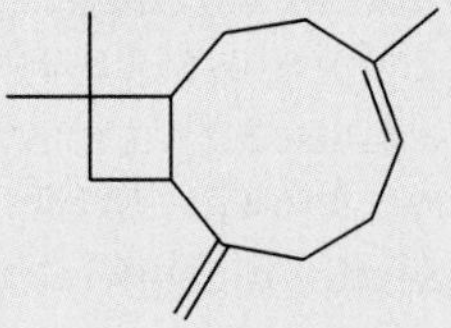

(c)

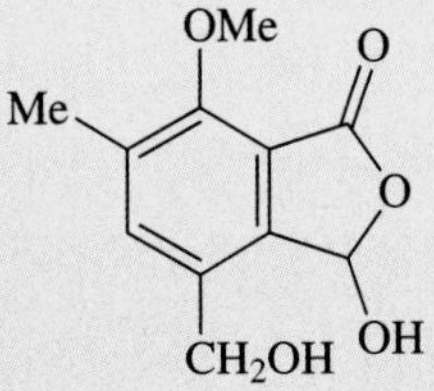

(d)

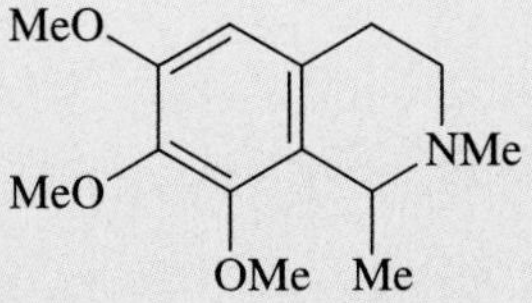

(e)   HO

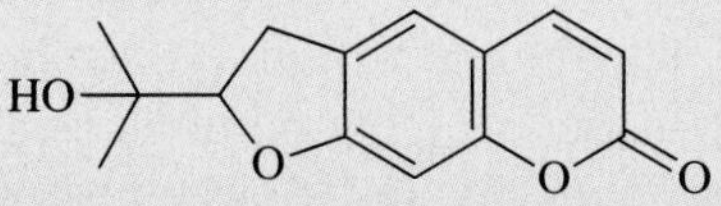

5.3. The following C_{10} compound has been isolated from a fungus. It may be a polyketide with the extra methyl groups introduced from the C_1 pool or it may be a monoterpenoid. Suggest a series of labelling experiments which might distinguish between the possibilities, giving the predicted results for each pathway.

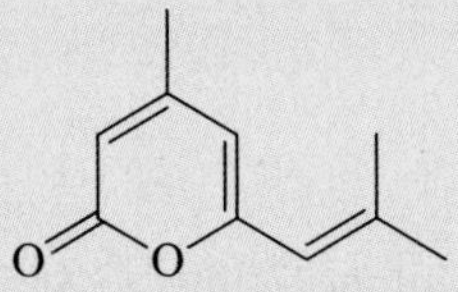

5.4. A plausible biosynthesis of the following natural product is from α-ketoglutaric acid [2-oxopentane-1,5-dioic acid, HO_2C–$C(O)CH_2CH_2$–CO_2H] and a fatty acid. Indicate the constituent units and show how they might be linked together.

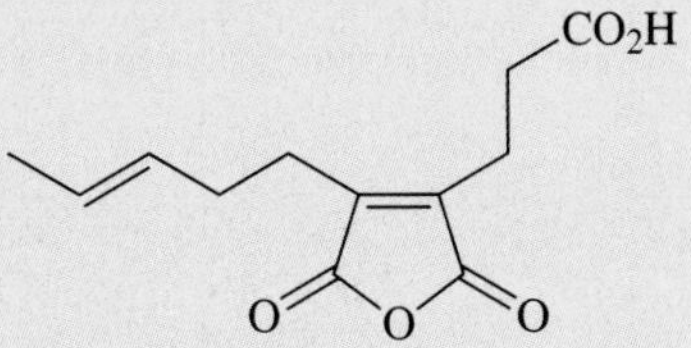

5.5. The carbon skeleton of the sesquiterpenoid fungal metabolite trichothecolone might be formed by folding the farnesyl pyrophosphate precursor in either of the modes **A** or **B**. When $[1,2\text{-}^{13}C_2]$acetic acid was used as a precursor, the trichothecolone had the coupling pattern shown. Which folding of farnesyl pyrophosphate is consistent with this labelling pattern?

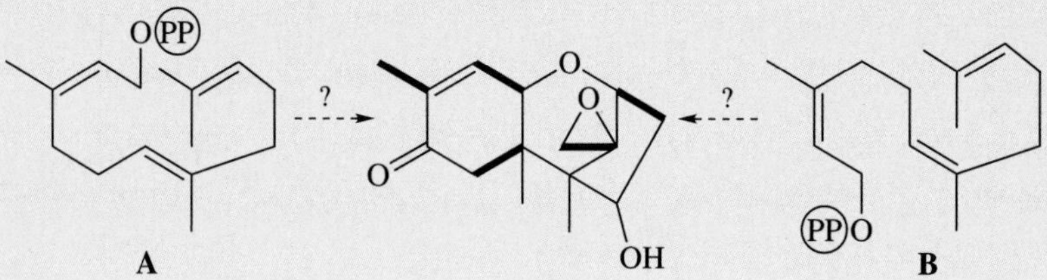

5.6. The fungal metabolite terrein (**C**) is biosynthesized from $[1,2\text{-}^{13}C_2]$acetate with the coupling patterns as shown. The carbon atoms that are enriched by $[1\text{-}^{13}C]$acetate are indicated by an asterisk. The isocoumarin **D** is a co-metabolite and is efficiently converted into terrein. Which carbon atoms are lost in this biosynthesis?

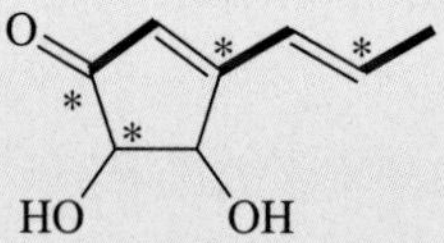

5.7. The biosynthesis of penicillin G from its constituent amino acids involves a tripeptide, L-aminoadipoyl-L-cysteinyl-D-valine, in which the valine has the unusual D configuration. The isomerization from L-valine may take place either before or after the tripeptide is formed. Suggest an experiment to distinguish between these possibilities.

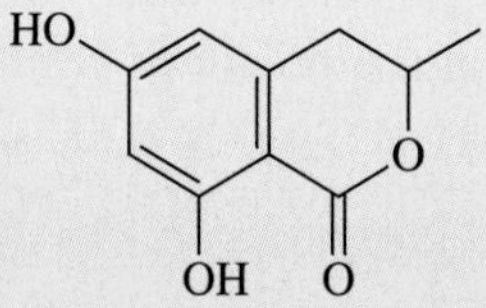

Further Reading

There is a substantial literature on natural products. Amongst the more detailed textbooks are *Medicinal Natural Products* by Dewick,[1] *The Chemistry of Natural Products* by Thomson,[2] and *Natural Product Chemistry* by Torsell[3] and *The Chemistry of Biomolecules* by Simmonds.[4] Textbooks entitled *Chemical Aspects of Biosynthesis* by Mann[5] and *The Biosynthesis of Secondary Metabolites* by Herbert,[6] as their titles suggest, cover the biosynthesis of natural products.

The series of volumes *Comprehensive Natural Product Chemistry*[7] covers the subject in considerable detail.

There are a number of reference books which cover specific areas of natural products, such as Simonsen's *Terpenoids*,[8] Manske's *The Alkaloids*,[9] Pelletier's *The Alkaloids*,[10] Fieser and Fieser's *Steroids*,[11] Harborne's *The Flavonoids*,[12] and *Fungal Metabolites* by Turner and Aldridge.[13] Methods for the extraction and identification of natural products are described in *Phytochemical Methods* by Harborne.[14] The biological activity of natural products is described in *Murder, Magic and Medicine* by Mann[15] and in an *Introduction to Ecological Biochemistry* by Harborne.[16]

There are a number of volumes which appear regularly with reviews on natural product chemistry, including *Progress in the Chemistry of Organic Natural Products*[17] and *Studies in Natural Product Chemistry*.[18] A number of reference books are devoted to natural products, giving structures, physical constants and sources, including the *Dictionary of Natural Products*,[19] *Dictionary of Terpenoids*[20] and the *Dictionary of Steroids*.[21] The single volume *Merck Index*[22] and the *ROMPP Encyclopedia of Natural Products*[23] are also useful sources of information.

The review journal *Natural Product Reports* contains regular annual or biennial articles covering particular groups of natural products, together with a large number of authoritative articles on topics in bio-organic chemistry. *Natural Product Updates* monitors a large number of journals for papers of interest to natural product chemists and provides a graphical abstract of each paper. Among the specialist journals devoted to natural products are *Phytochemistry*, the *Journal of Natural Products*, the *Journal of Antibiotics*, *Natural Product Letters*, the *Journal of*

Chemical Ecology, *Steroids*, and *Phytochemical Methods*. Other articles appear regularly in the journals devoted to organic chemistry, such as the *Journal of the Chemical Society, Perkin Transactions 1* and its successor, *Organic and Biomolecular Chemistry*, the *Journal of Organic Chemistry*, *Chemical and Pharmaceutical Bulletin*, *Tetrahedron*, *Tetrahedron Asymmetry* and *Tetrahedron Letters*.

References

1. P. M. Dewick, *Medicinal Natural Products – A Biosynthetic Approach*, 2nd edn., Wiley, Chichester, 2000.
2. R. H. Thomson (ed.), *The Chemistry of Natural Products*, 2nd edn., Blackie, London, 1993.
3. K. B. G. Torsell, *Natural Product Chemistry*, 2nd edn., Apotekarsocieteten, Stockholm, 1997.
4. R. J. Simmonds, *Chemistry of Biomolecules*, Royal Society of Chemistry, Cambridge, 1992.
5. J. Mann, *Chemical Aspects of Biosynthesis*, Oxford University Press, Oxford, 1994.
6. R. B. Herbert, *The Biosynthesis of Secondary Metabolites*, Chapman and Hall, London, 1981.
7. D. H. R. Barton and K. Nakanishi (eds.), *Comprehensive Natural Product Chemistry*, Elsevier, Amsterdam, 1999, vols. 1–9.
8. J. Simonsen, D. H. R. Barton and L. N. Owen (eds.), *The Terpenoids*, 2nd edn., Cambridge University Press, London, 1961, vols. 1–5.
9. R. H. F. Manske *et al.* (eds.), *The Alkaloids*, Academic Press, New York, 1953–2000, vols. 1–53.
10. S. W. Pelletier (ed.), *The Alkaloids, Chemical and Biological Perspectives*, Elsevier, Amsterdam, 1999, vols. 1–14.
11. L. F. Fieser and M. Fieser, *Steroids*, Reinhold, New York, 1959.
12. J. B. Harborne and H. Baxter, *Handbook of Natural Flavonoids*, Wiley, Chichester, 1999, vols. 1 and 2.
13. W. B. Turner and D. C. Aldridge, *Fungal Metabolites II*, Academic Press, London, 1983.
14. J. B. Harborne, *Phytochemical Methods*, 2nd edn., Chapman and Hall, London, 1984.
15. J. Mann, *Murder, Magic and Medicine*, Oxford University Press, Oxford, 1992.
16. J. B. Harborne, *Introduction to Ecological Biochemistry*, 4th edn., Academic Press, London, 1993.

17. L. Zechmeister *et al.* (eds.), *Progress in the Chemistry of Organic Natural Products*, Springer, Berlin, 1938–2000, vols. 1–79.
18. Atta-ur-Rahman (ed.), *Studies in Natural Product Chemistry*, Elsevier, Amsterdam, 1999, vols. 1–20.
19. J. Buckingham (ed.), *Dictionary of Natural Products*, Chapman and Hall, London, 1994.
20. J. Connolly and R. Hill (eds.), *Dictionary of Terpenoids*, Chapman and Hall, London, 1991.
21. D. N. Kirk (ed.), *Dictionary of Steroids*, Chapman and Hall, London, 1992.
22. *The Merck Index*, 12th edn., Merck, New York, 1996.
23. W. Steglich, B. Fugmann and S. Lang-Fugmann (eds.), *ROMPP Encyclopedia of Natural Products*, Thieme, Stuttgart, 2000.

Answers to Problems

Chapter 1

1.1 (a) Polyketide; (b) phenylpropanoid; (c) polyketide; (d) terpene; (e) alkaloid; (f) phenylpropanoid; (g) sugar; (h) terpene; (i) alkaloid.

1.2.

(a)

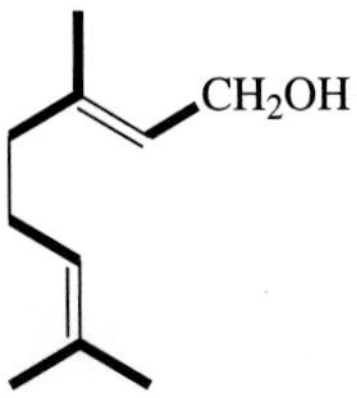

CH₂OH

(b) CH₂OH

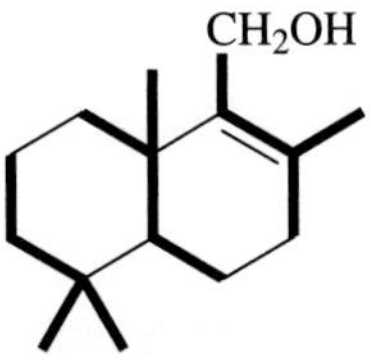

(c) CO₂H

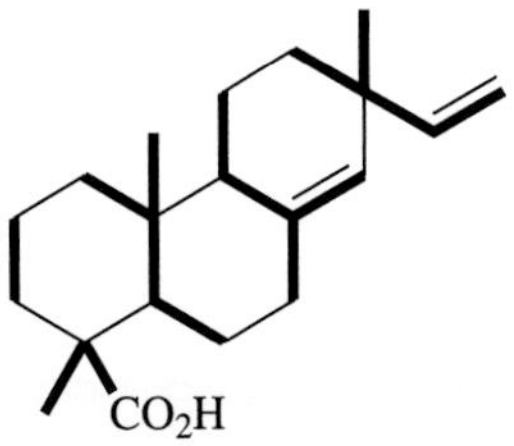

1.3.

(a) HO / OH / O

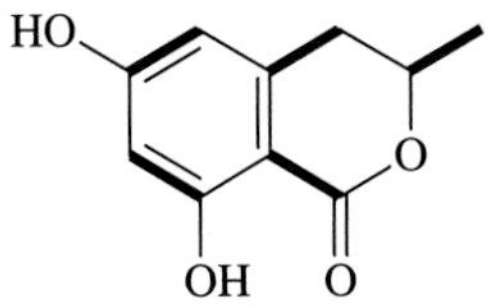

(b) HO / O / O / OH / OH

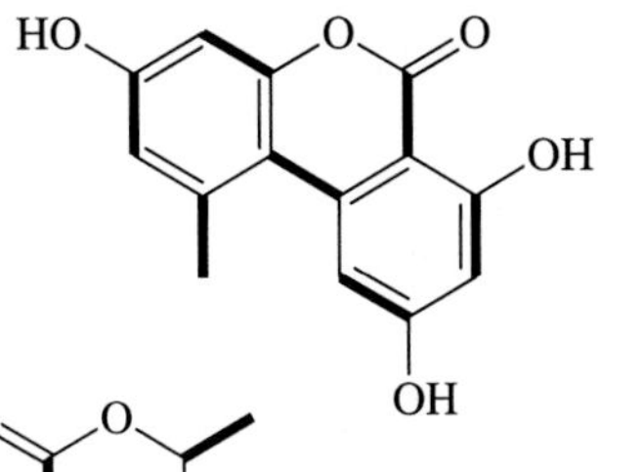

(c) HO / OH / O / O / O

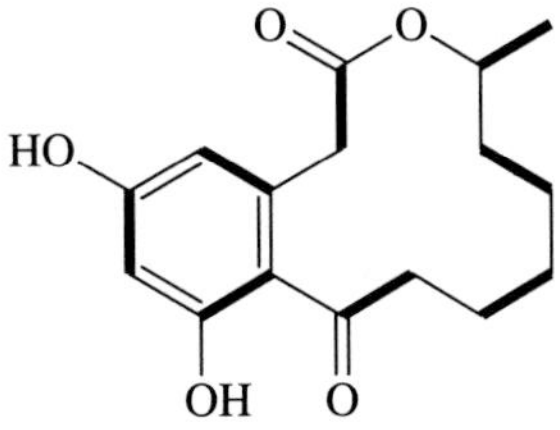

1.4.

Dissolve extract in ethyl acetate

Extract with dil. HCl   Neutral and acidic material

Alkaloids dissolve as salts

Make basic with ammonia

Recover alkaloids in ethyl acetate

1.5. The lower part of the molecule represents the possible pharmacophore:

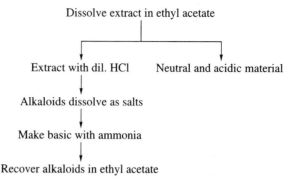

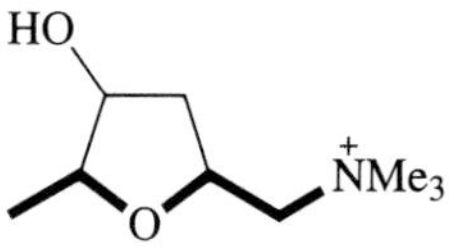

Chapter 2

2.1. (a) (i) The ^{1}H NMR spectrum of compound **A** will contain three alkene and two methyl group resonances while **B** will contain one alkene and three methyl group resonances. (ii) Compound **A** will give formaldehyde on ozonolysis.

(b) (i) The UV spectrum of compound **D** will show the presence of a longer chromophore; or the ^{1}H NMR spectrum of compound **D** will contain a methyl group doublet. (ii) There is only one exchangeable proton in compound **D** when the compound is treated with NaOD/ D_2O; or compound **D** contains a diene which could undergo a Diels–Alder addition.

(c) (i) The ^{1}H NMR spectrum of compound **E** will contain a methyl group singlet and a CH_2O resonance, in contrast to compound **F**. The ^{13}C NMR spectrum of compound **E** will contain a tertiary alcohol resonance. (ii) Oxidation of compound **F** will give a cyclic anhydride.

(d) (i) Compound **G** is symmetrical and this will be shown by the ^{1}H and ^{13}C NMR spectra, whereas compound **H** will show separate CH_2OH, CH_2OAc and CHOH signals. (ii) Compound **H** is a 1,2-diol and will be oxidized by sodium iodate(VII).

(e) (i) The IR spectrum of compound **J** will show the carbonyl absorption of an amide whilst that of compound **I** will be a carboxylic acid. (ii) Compound **I** will dissolve in sodium hydrogen carbonate as an acid, while compound **J** will release ammonia on alkaline hydrolysis.

(f) (i) The ^{1}H NMR spectrum of compound **K** will contain a CH(OH), a CH_2OH and five aromatic proton resonances, but that of compound **L** will contain four aromatic proton resonances and a CH_2OH signal; or compound **L** will behave as a phenol, and in the UV spectrum there will be a change in the position of λ_{max} on addition of alkali. (ii) Oxidation of compound **K** will give a β-keto acid which will readily undergo decarboxylation.

2.2. (a) The molecular formula $C_8H_{14}O$ corresponds to two double bond equivalents. (b) The compound contains a carbonyl group. (c) The compound contains three isolated methyl groups adjacent to unsaturation and a trisubstituted double bond. (d) The formation of propanone indicates the presence of the group:

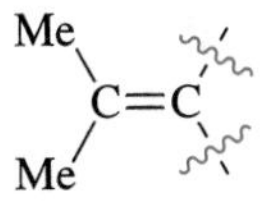

The formation of triiodomethane (iodoform) indicates the presence of a methyl ketone ($CH_3C=O$). Therefore the structure of compound **M** is:

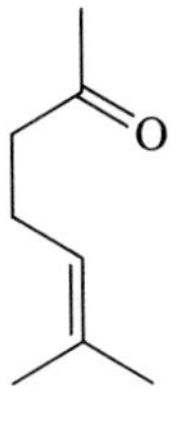

2.3. Compound **N** is:

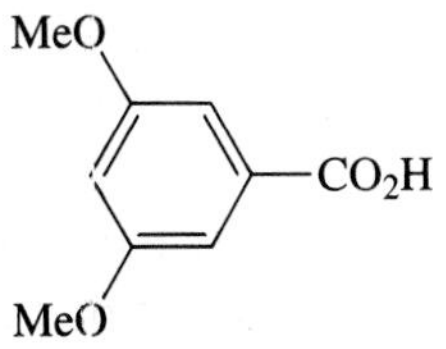

The small couplings are long-range (*meta*) couplings.

2.4.   Compound **O** is:

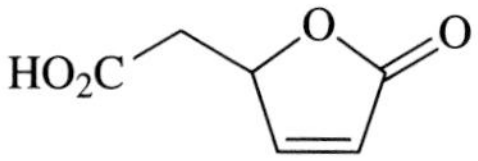

Compound **O** contains a lactone ring which undergoes hydrolysis in alkali to the corresponding hydroxy acid (**P**).

2.5.   Sherry lactone and solerone are:

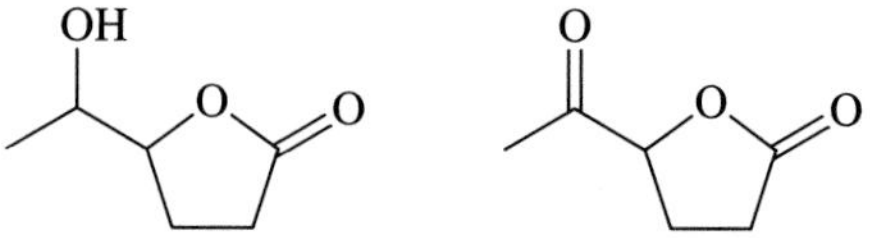

2.6.   The sesquiterpenoid **Q** is:

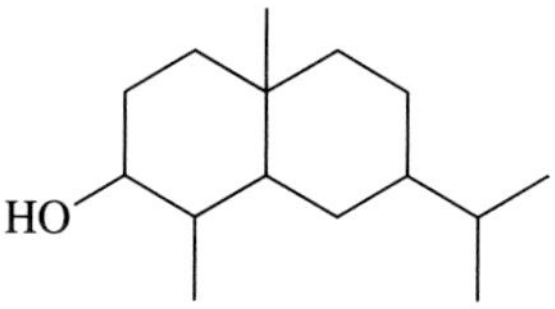

There are two possible dissections into the constituent isoprene units:

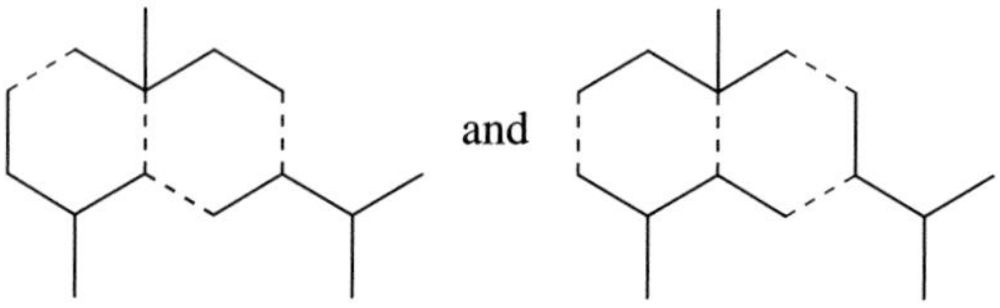

After you have read Chapter 5, consider how you might distinguish between them.

2.7.   Compounds **R** and **S** are:

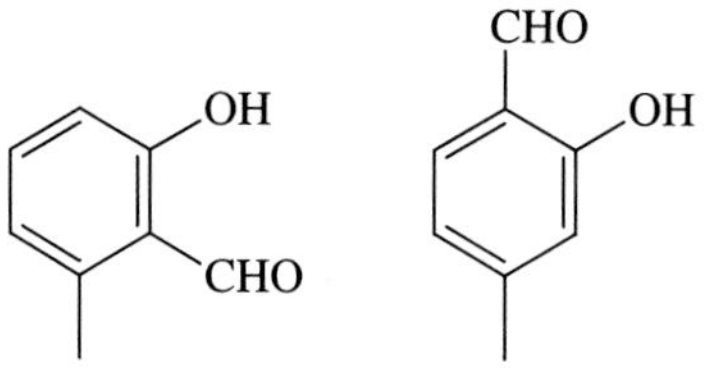

R   S

2.8. Compound **T** is:

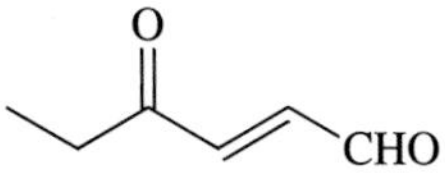

2.9. Compounds **U**, **V**, **W** and **X** are:

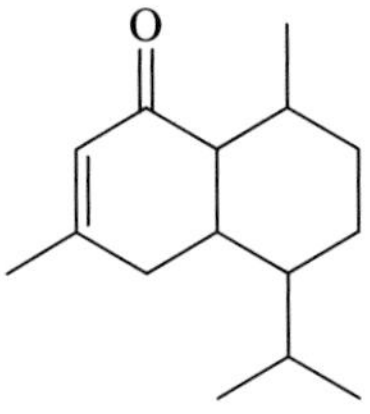

U

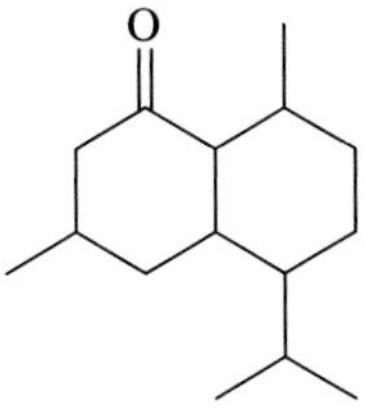

V

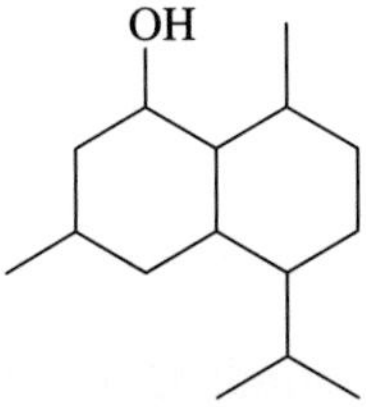

W

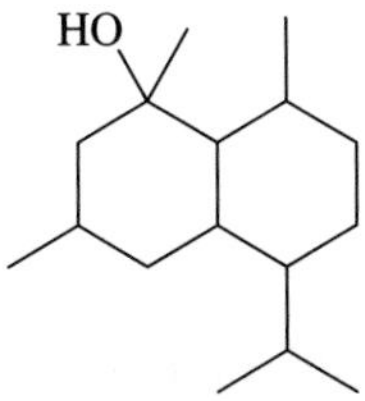

X

Chapter 3

3.1. Compounds **A**, **B** and **C** are:

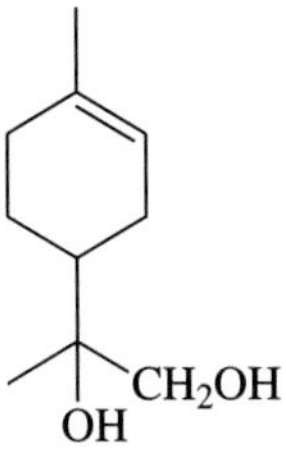

A

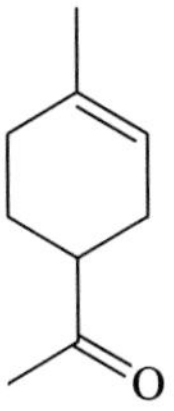

B

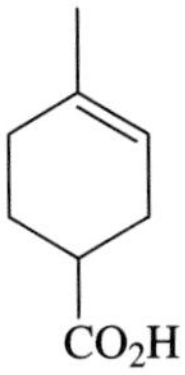

C

3.2. The stereochemistry of 5-deoxypulchelloside is:

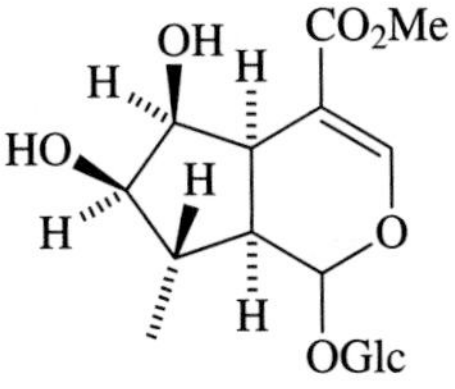

3.3.   The C-6 hydroxyl group which is involved in lactone formation must lie on the same (β) face of the molecule as the nitrogen bridge. The other hydroxyl group at C-3 is α.

3.4.   Alkaloid **G** is:

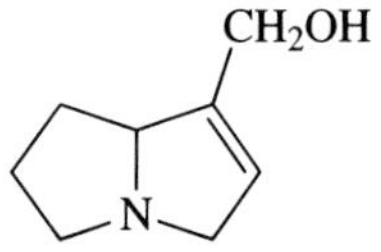

3.5.   Alkaloid **K** is:

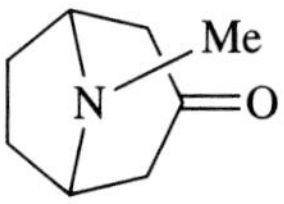

3.6.   The formation of the acetonide from the secondary alcohols of ribose indicates that they are *cis* to each other and *trans* to the primary alcohol. The formation of the cyclic salt can only occur if the primary alcohol of the D-ribose and the anomeric C–N bond are *cis*.

3.7.   The antibiotic **P** and the compounds **Q** and **R** are:

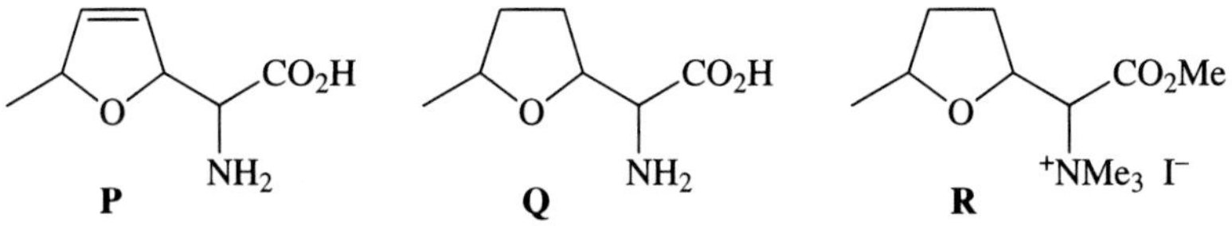

P   **Q**   **R**

3.8.   The fungal metabolite **S** is:

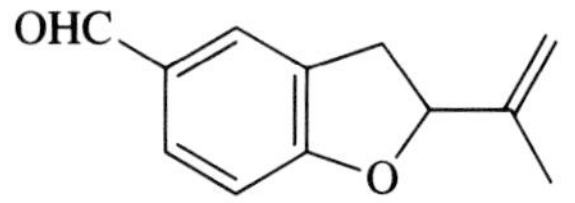

4.1.   Apional is:

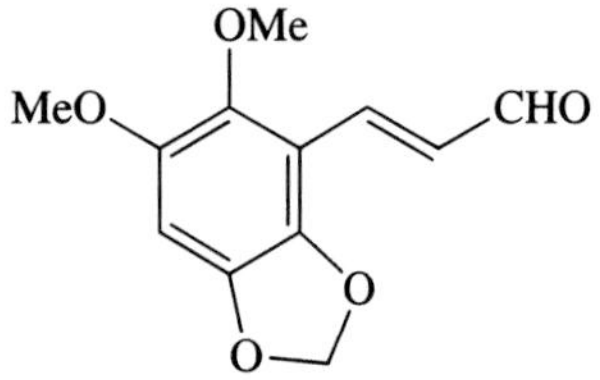

4.2.   Compounds **A** and **B** are:

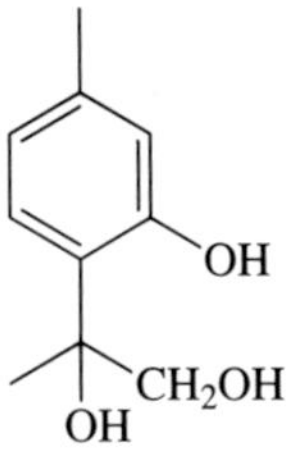

OH

CH₂OH
OH

A

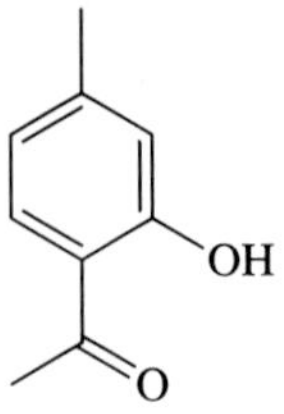

OH

O

B

4.3.   Trachelantic acid (compound **C**) is:

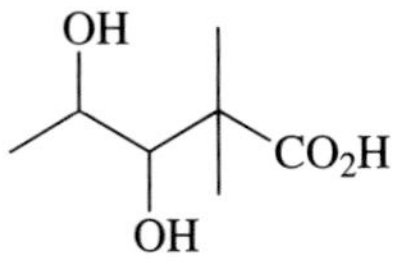

OH

CO₂H

OH

4.4.   Terrein is:

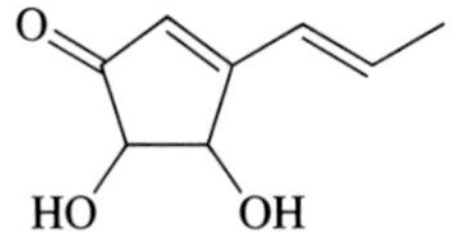

O

HO   OH

4.5.   Paeonol is:

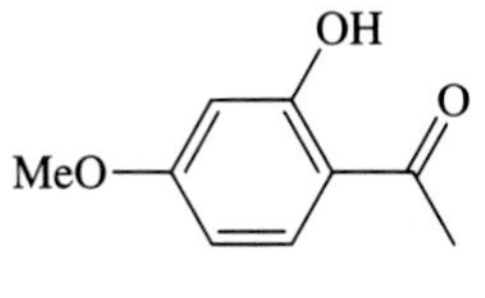

OH

O

MeO

4.6.   Pyrenocine **C** is:

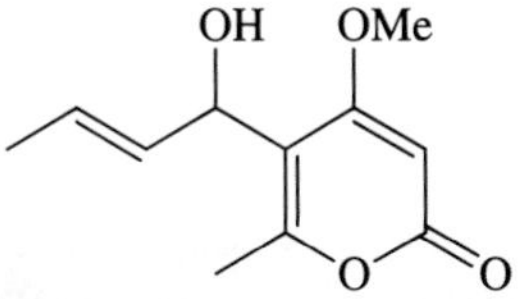

OH   OMe

O   O

4.7.   Compound **G** is:

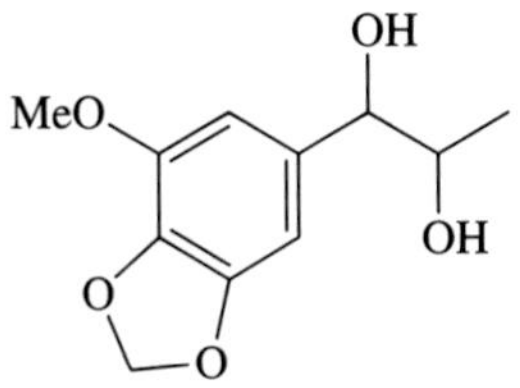

OH

MeO

OH

O

O

Chapter 5

5.1.

(a)

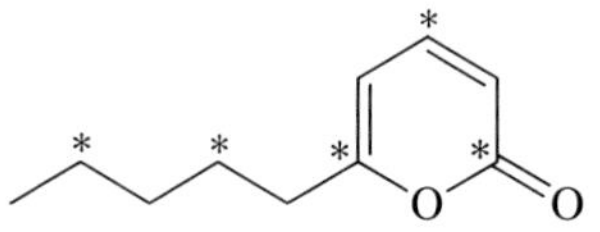

(b)

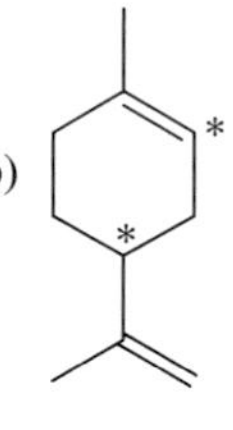

(c)

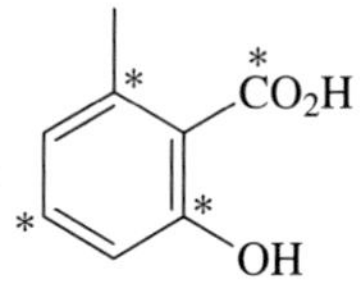

(d)

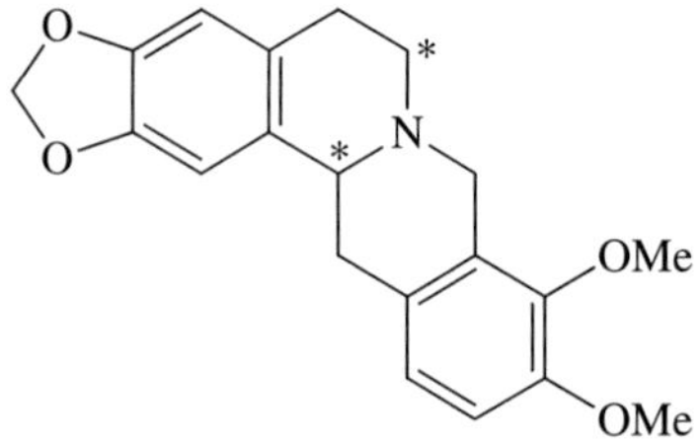

(e)

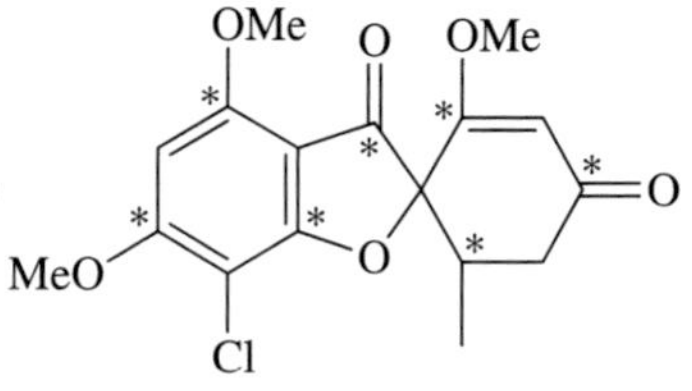

5.2.

(a)

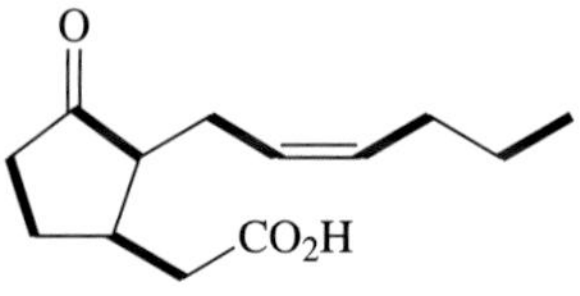

acetate

(b)

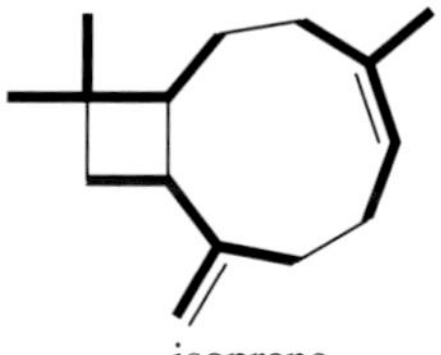

isoprene

(c)

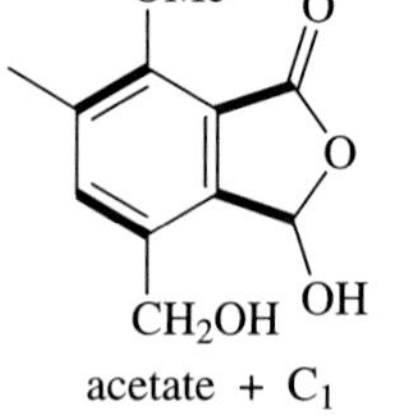

acetate + C_1

(d)

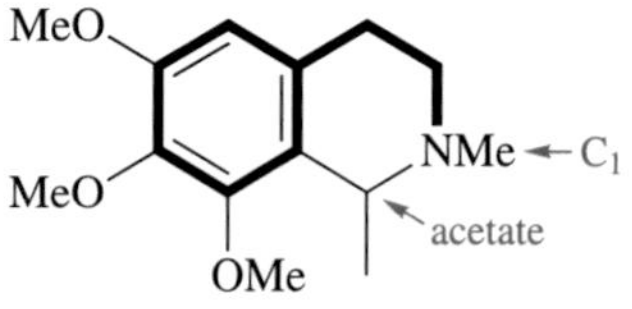

C_6–C_3 via tyrosine

(e)

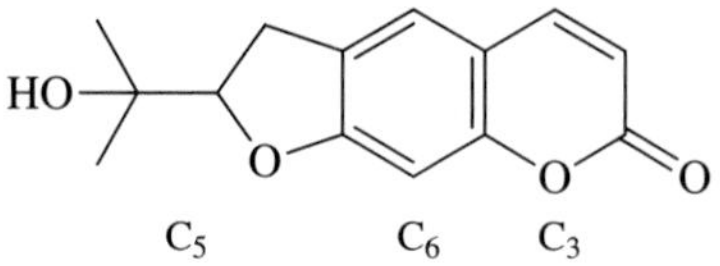

C_5 C_6 C_3

C_6–C_3 + isoprene

5.3. Use $[1,2-^{13}C_2]$acetate and methyl-labelled methionine. The methyl groups will either be labelled by acetate or from methionine (the biosynthesis actually takes place via the polyketide pathway and extra methylations).

5.4. The constituent units are:

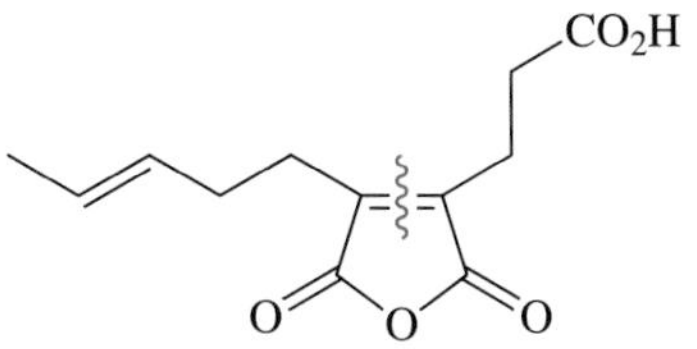

The association of the two units probably takes place by a carbanion process.

5.5. Farnesyl pyrophosphate is labelled by $[1,2-^{13}C_2]$acetate as shown, and thus folding **A** is consistent with this labelling pattern.

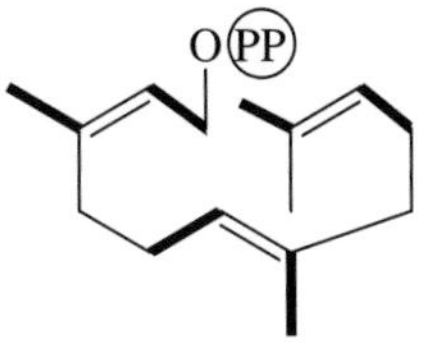

5.6. The isocoumarin **D** will be labelled by the acetate units as shown:

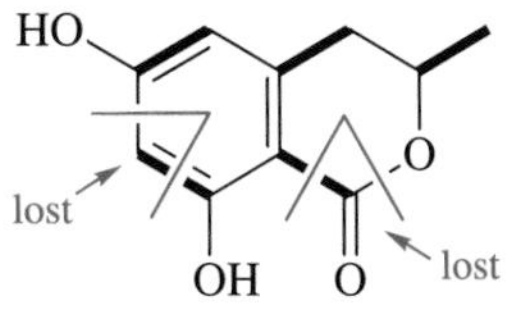

Consequently the atoms that are lost are those shown.

5.7. The experiment might involve feeding a sample of DL-valine, in which one enantiomer is labelled with carbon-14 and the other with tritium. If the isomerization takes place prior to incorporation, both isotopes will be incorporated, whereas if it is after the formation of the tripeptide, only one isotope will be incorporated. In practice the isomerization takes place after the tripeptide is formed.

Subject Index

Abietic acid 9, 51
Abscisic acid 8
Absolute stereo-
 chemistry 75
Acetonide
 formation 72
Acetyl co-enzyme A
 108
Acid-catalysed
 degradation 56
Aflatoxins 5
Aglycone 3
Alkaline
 degradation 55
Alkaloids 2, 18, 29,
 121
Alkenes (IR, NMR) 40
Allelopathy 25
Amino acids 2, 18, 2,
 121, 124
Amyrins 11
Anthocyanins 16
Apoaromadendrene
 69
Arachidonic acid 4,
 109
Arecoline 18
Artemisinin 8
Aspyrone 111
Asymmetric
 induction 77
Atrolactic acid 77
Atropine 18
Auxochrome 43
Azadirachtin 12

Berberine 19
Biosynthetic
 methodology
 105–106

Biotin 32, 108
Botrydial 9

Caffeine 21
Camphor 7
Carbon-13 NMR
 spectroscopy 44, 111
Carbon skeleton,
 determination 46
Carbonyl groups (IR,
 NMR) 40
β-Carotene 14
Carotenoids 7, 14
Carvone 75
Caryophyllene 8
Cedrene 8
Cephalosporin C 95
Characterization,
 natural products 35
Chemical
 ionization 37
Chlorophyll 125
Cholesterol 12, 116
Chorismic acid 120
Chrysin 17
Circular dichroism 78
Clavatol 70
Clavulanic acid 95
Clerodanes 10
Cocaine 18
Coniine 18
Cortisone 13
COSY 45, 58
Cotton effect 77
Coupling, long-range
 67

Degradation
 acid-catalysed 56
 alkaline 56

oxidative 47
 reductive 50
Dehydrogenation 50
Dehydrogeosmin 38
1-Deoxyxylulose
 pathway 113
DEPT 44, 58, 65
Determination, carbon
 skeleton 46
Dimethylallyl
 pyrophosphate 113
Diosgenin 13
Diterpenoids 6
Double bond
 equivalents 36

Ecdysone 27
Electron impact
 ionization 37
Ephedrine 19
Ergosterol 42
Erythromycin 4
Estradiol 12, 116
Eugenol 15
Exobrevicomin 27

Farnesyl
 pyrophosphate 114
Fast atom
 bombardment 38
Fatty acids 3
Fischer 76
Flavanone 16, 120
Folic acid 23

(+)-Glyceraldehyde
 76
Genistein 17
Geraniol 7

Geranyl pyrophosphate 113
Gibberellic acid 10, 51, 56, 117
Gibberellin A$_{13}$ 71, 73
Gibberic acid 69
Gliotoxin 22
Glucose 72
Glycosides 3
Griseofulvin 5, 43, 89–92, 111

Heptulosinic acid 119
Hofmann elimination 53
Horeau's method 77
Humulene 8
Hydroxy groups 39

Indole alkaloids 20
Induction, asymmetric 77
Infrared spectroscopy 38, 68
Insect juvenile hormone 27
Ipsdienol 27
Iresin 73
Iridoids 8
Isopentenyl pyrophosphate 6, 112
Isoprene rule 6

Jasmonic acid 4
Juglone 25

Karplus equation 66
ent-Kaurene 10, 117
Ketomanoyl oxide 52, 69
Kuhn–Roth determination 47

Lignin 17
Limonin 11
Linalool 7
Linoleic acid 3
Linolenic acid 3
Long-range coupling 67

Longifolene 8
Luteolin 56

Macrolides 4
Manoyl oxide 9, 52
Mass spectrometry 37
Mellein 5, 111
Menthol 7
Metabolites
    primary 1
    secondary 2
Methyl podocarpate 75
6-Methylsalicylic acid 110
Methyl vinhaticoate 75
Mevalonic acid 112
Microhydrogenation 46
Molecular rotation differences 76
Monosaccharides 76
Monoterpenoids 6
Morphine 20, 123
MTPA esters 79
Mycophenolic acid 5
Myrtenol 50

Narigenin 16
Natural products, characterization 35
Nicotine 19, 26, 122
NMR spectroscopy 43, 57, 64–65, 79
Nuclear Overhauser effect 58, 66

Octant rule 78
Oleic acid 3
Optical rotatory dispersion 77
Oxidative degradation 47
Ozonolysis 72, 96

Papaverine 19, 47, 122
Patulin 5
Penicillin 22, 93–95, 124–125

Phaseolin 25
Phenylpropanoids 2, 15, 119–121
Phytoalexins 25
Pimaric acid 9
α-Pinene 7, 36, 48
β-Pinene 49
Piperine 18
Podophyllotoxin 15
Polyketide synthesis 108–109
Polyketides 2, 3–6, 108–111
Prephenic acid 120
Primary metabolites 1
Progesterone 12, 116
Prostaglandins 96–98, 110
Pseudopelletierine 54
Pyrethrins 26

Quinine 20

Reductive degradation 50
Reticuline 123
Rishitin 25

Salutaridinone 123
Santonin 8, 56, 85–89
Scopine 72
Secondary metabolites 2
Sesquiterpenoids 6
Shikimic acid 15, 119
Sodium iodate(VII) 70
Squalene 11, 114
Stearic acid 3
Stereochemistry, absolute 75
Steviol 10, 74
Strychnine 21

Taxol 10
Terpenoids 2, 6–14, 112
Terpineol 48
Testosterone 13, 116
Tetracyclin 5
Trichothecin 9
Triterpenoids 7
Tryptophan 121

Ultraviolet
    spectroscopy   41
Umbelliferone   15
Uro'gen III   125

Verbenol   27
Vitamin A   14, 23
Vitamin B_1 23, 42

Vitamin B_2 23, 125
Vitamin B_6 23
Vitamin B_{12} 23, 125
Vitamin C   98–101
Vitamin D   13
Vitamin E   24
von Braun
    degradation   55

Wagner–Meerwein
    rearrangement   56,
    74
Woodward–Fieser
    rules   41
Wyerone   4